Oxygen Diving

The book provides a derivation of the models used for calculating the risk and hazard of central oxygen toxicity pertaining to diving-based studies consistent with the research conducted earlier by the Royal Navy and the US Navy. This book forms the basis for extending the possibility of undertaking nitrox dives in combination with oxygen dives, thus significantly increasing tactical capabilities of conducting diving special operations.

Features:

- provides derivation of the models used for calculating the risk and hazard of central oxygen toxicity
- improves oxygen diving procedures described in the US Navy Diving Manual
- includes procedures applicable to undertaking nitrox dives in combination with oxygen dives
- pitches the material at highest technology readiness levels, i.e. 9 TRL
- aims to increase tactical capabilities of conducting diving special operations

This book is aimed at researchers, professionals and graduate students in life support system design, diving submarine safety, ventilation, health, sanitary engineering, mining engineering and working environment in chambers or closed compartments.

Diving Sciences

Series Editor: Ryszard Kłos

The series is aimed at encouraging scientists from around the world to publish research related to diving and hyperbaric exposures pertaining to the last Technology Readiness Levels (8 TRL-9 TRL), with a particular focus on system prototype demonstration in an operational environment, system complete and qualified, and actual system proven in operational environment. The intended audience includes researchers in diving science, military and civilian divers, diving contractors, navies, defense and civilian diving schools and related academic audiences.

Oxygen Diving
Ryszard Kłos

For more information about this series, please visit: https://www.routledge.com/Diving-Sciences/book-series/CRCDS

Oxygen Diving

Ryszard Kłos

CRC Press
Taylor & Francis Group
Boca Raton London New York

CRC Press is an imprint of the
Taylor & Francis Group, an **informa** business

First edition published 2023
by CRC Press
6000 Broken Sound Parkway NW, Suite 300, Boca Raton, FL 33487-2742

and by CRC Press
4 Park Square, Milton Park, Abingdon, Oxon, OX14 4RN

CRC Press is an imprint of Taylor & Francis Group, LLC

Reasonable efforts have been made to publish reliable data and information, but the author and publisher cannot assume responsibility for the validity of all materials or the consequences of their use. The authors and publishers have attempted to trace the copyright holders of all material reproduced in this publication and apologize to copyright holders if permission to publish in this form has not been obtained. If any copyright material has not been acknowledged please write and let us know so we may rectify in any future reprint.

Trademark notice: Product or corporate names may be trademarks or registered trademarks and are used only for identification and explanation without intent to infringe.

ISBN: 978-1-032-31389-4 (hbk)
ISBN: 978-1-032-31390-0 (pbk)
ISBN: 978-1-003-30950-5 (ebk)

DOI: 10.1201/9781003309505

Typeset in Times
by Apex CoVantage, LLC

In appreciation of her assistance and perseverance, I dedicate this book to my dear wife Yvonne, who through the years has supported and aided me in my efforts.

Contents

About the Author

Ryszard Kłos graduated from the Technical University of Wrocław with an MSc in physicochemistry. He started his professional carrier in the Institute of Low Temperature and Structural Research, Polish Academy of Sciences, Wrocław.

In 1985 he attended military training required of university graduates and graduated from Chemical Warfare College. Next year he joined the Polish Navy as platoon leader in Military Diving School. In 1988 he was transferred to the Naval Academy to the Department of Diving Gear and Underwater Work Technology where he earned his PhD in the field of design and operation of machines. He completed a post-doctoral fellowship at the DR-DC Toronto (former: Defence and Civil Institute of Environmental Medicine). He defended his habilitation thesis (higher PhD thesis) in the field of machine operation at Gdańsk University of Technology.

The author is a retired Polish Navy captain. He is currently employed as Associate Professor at the Polish Naval Academy, where he is responsible for various diving and physicochemical research projects and programs. He has written nine books and many scientific articles. From 2012 to 2015 he served as Deputy of Rector for Scientific Research at the Polish Naval Academy. From 2004 to 2007 he was President of the Polish Hyperbaric Medical and Technical Society.

Preface

The book is one of the publications in the series Diving Sciences aimed at publishing research on diving and hyperbaric exposures pertaining to technology readiness levels (TRL). It is especially focused on a system prototype demonstration in an operational environment, a system complete and qualified, and a system proven in an operational environment. The intended audience includes researchers in diving science, military and civilian divers, and diving contractors, navies, defense and civilian diving schools and related academic audiences.

This book will be of use to diving scientists, as they must assess the risk of central nervous syndrome, and it will also be useful for special operations forces tactics specialists, because:

- navies expect progress in technologies of diving dedicated to use in military operations
- defense diving schools expect materials for training in modernized technologies of diving
- military divers expect aided materials in the process of self-education

Civilian diving with oxygen is illegal in many countries, and its use is only possible during the decompression phase, where the risk of oxygen toxicity is minimal. The models described in the book allow assessing the risk of oxygen toxicity, which may be useful for technical divers.

The book deals with the validation process of a real system proven in an operational environment and is not intended as a summary or scientific discussion of current trends in biochemical advances in oxygen toxicity research. The book explains only the necessary biological mechanisms needed to understand the content, bearing in mind that it will be read by users who do not need specialist knowledge of biology, chemistry, biochemistry, biophysics, toxicology and so on, and focuses only on oxygen toxicity models that have been proven beyond any doubt with an assumed accuracy limit needed in practical use. This validation process is described in detail in the book as its aim/canvas.

Validation studies are time-consuming and require a high budget. If someone would like to save a lot of time and money, it is much better to first evaluate the scientific process carried out in Poland and then implement their own research.

The described model of central nervous syndrome caused by oxygen toxicity is based on survival analysis. It was based on earlier experiments performed by Great Britain during World War II and introduced by the United States in the 1970s. This theory is still valid and is used by most navies.

During the research conducted in Poland the model was validated, which is described in the book. Additionally, this technology was further developed and now embraces not only the oxygen technology but the oxygen/nitrox one as well.

The approach is similar to finding a new drug therapy. The studies described in the book relate to clinical trials that were organized similarly to the clinical trials of new drug therapies with the approval of the Bioethics Committee in accordance with the Declaration of Helsinki. The research involved not the use of a new drug but the optimization of the procedure for its use. That means the research described in the book was formally carried out as the implementation of a new treatment procedure using a classic drug, which consisted of checking the optimal method of its dosing. Therefore, the book does not provide a critical analysis of current medical reports on the mechanism of the toxic effect of oxygen as its content relates to the possibility of practical optimal application of the achievements to date in this field.

An example of the equipment described in the book is a French-made diving apparatus, but all special operations forces can use the technology modified in Poland, especially if they use alternatively supplied diving apparatuses, such as

- *oxy – CCR/Nx – SCR AMPHORA SCUBA*[1]
- *oxy – CCR/Nx – SCR LAR VII Combi FA (LEBA 24/LAR 5010) SCUBA*
- *oxy – CCR/Nx – SCR Oxy – Mix SCUBA* and similar construction

but other countries can use this technology to a limited extent.

From the date of its establishment (1976), evaluation and study of diving equipment and decompression systems has been the primary task carried out by the Department of Underwater Work Technology[2] of the Naval Academy. The projects referred to required proper research stations. We mostly built them ourselves or acquired them special through orders. At the very beginning, the special purpose team of technicians was established for this task. They were able to produce technology demonstrators[3] as well as prototypes of diving and scientific instruments. Experimental Deepwater Diving Complex DGKN-120 was supposed to be a universal test bench and was designed to provide comprehensive research capabilities for the diving apparatus. At that time, the initial concepts of research stations were designed by the first head, late Cpt(N) Medard Przylipiak MSc Eng. The ideas of conducting a comprehensive study of diving apparatus were common with his successors: the late Cmdr. Marian Pleszewski MSc Eng. and Cpt(N) Stanislaw Skrzyński PhD Eng., which did not prevent them from making other scientific achievements.[4] One person who played a leading role was Cpt(N) Prof. Tadeusz Doboszyński PhD, DSc Med. from the Military Medical Academy.

It was Cpt(N) Stanislaw Skrzyński MSc Eng. who became an inspiring leader in the development and research on new types of diving equipment. The results of this work have been largely wasted due to the failure to promote research focused on developing the original Polish diving apparatus and decompression systems.

Bearing in mind the old concepts and the undeniable achievements, efforts were made to raise funds to continue the work done by their predecessors. Initially, the purchases of foreign diving equipment facilitated this task, at the same time creating a need to start the implementation work.[5] The effective stage of research was reached with modelling breathing in *CCR* and *SCR* systems, which made it possible to connect the diving apparatus model with the model of decompression diving. Another completed project made it possible to develop a statistical approach to the safety of diving. The research into saturation diving and decompression systems allowed us to extend our laboratory facilities and expand the filed specific knowledge. Using predecessors' ideas and not-so-small funds, we managed to build a material base and theoretical knowledge to carry out research on diving apparatuses and decompression systems. The research facilities are also used for training and as an emergency center. It is hoped that these efforts will not be wasted and Poland will return to and continue to be present among the countries developing diving technologies.

NOTES

1 *CCR* – closed circuit rebreather; *SCR* – semiclosed circuit rebreather; *Nx* – nitrox; *SCUBA* – self-contained breathing apparatus.
2 Formerly the Department of Diving Gear and Underwater Work Technology.
3 By technology demonstrator it is meant the practical demonstration of technical solutions aimed at the development of a particular system, demonstrating the possibility of achieving the required parameters for this system.
4 For example, the implementation of saturation dives.
5 But giving only evidence of the possibility of using domestic supplies.

Acknowledgments

Apart from financing, the presented research required access to unique combat equipment. I express my gratitude to the Navy Command of the Republic of Poland and the 3rd Ship Flotilla for the trust that I received.

The research included experiments on humans, which required consent of the Scientific Research Ethics Committee. In addition to appreciation to the predecessors, gratitude must be expressed for the contribution given by the research staff of the department and cooperating medical doctors from the Military Medical Institute. Of them Cmdr. Maciej Konarski PhD Med. has significantly contributed to assembling the experimental data described in this monograph.

Thanks for the confidence demonstrated by the divers and professionals from military units, police, anti-terrorist groups and others who have supported this work. Also, because of the considerable investment by the Ministry of Defense, Ministry of Science and Higher Education, and National Research and Development in the research carried out in the department, the desired and expected results have been attained.

In particular, I would like to express my gratitude to Karina Kowalska MSc, Prof. Grzegorz Kowalski PhD and Kazimierz Szczepański PhD, who reviewed the entire manuscript.

Definitions

Item	Description
Bathynautics	is the totality of knowledge about human underwater activities.
Bradycardia	is herein defined as the state when the heart rate decreases upon exposure of the body to hyperbaric conditions compared to the heart in normobaric conditions.
Buffer	is a solution whose *pH* value after the addition of small amounts of strong acids or alkali as well as after dilution with water is relatively constant.
Catalysts	are substances that increase the rate of reaction but remain unchanged after the reaction.
Catalytic reaction	(catalysis) is a chemical reaction in which a change in reaction rate occurs under the influence of a catalyst
Chemical equation	(reaction equation) is a quick record of the chemical reaction; the starting materials (substrates) are located on the left side of the formula and the substances obtained in the reaction (reaction products) are on the right side.
Chemical reaction yield	(efficiency of a chemical reaction) is the ratio of the amount of reacted material to the amount of matter that could theoretically react by chemical reaction.
Cycle ergometer	is a device with a structure similar to a bicycle, controlled by inhibiting the pedal effort at a controlled rate of the cycling.
Cytochromes	proteins present in the cell mitochondria with biocatalyst function involved in electron transport.
Decompression	with respect to the dive is a gradual lowering of pressure applied to the diver at the exit of the hyperbaric environment carried out so that there are no signs of disease.
Diffusion	is the spontaneous mixing of the components of the system as a result of the chaotic motion of particles.
Dissimulation	is understood here as a presence manifested in an attempt to meet the norms or hide the actual situation and feelings, impulses, behaviors, etc.
Endothermic reaction	is chemical reaction that occurs with the heat being absorbed.
Enzyme	is mostly protein and macromolecular chemical compound regulating life processes.
Equilibrium constant[1]	is the number given by the ratio of the product of the molar concentrations of the products to the product of the molar concentration of substrates – the molar concentrations should be expressed in molar fractions and raised to the power such as results from the reaction stoichiometric ratio
Exothermic reaction	is a chemical reaction that proceeds with the liberation of heat.
Homeostasis	is the ability of a living organism, by proper coordination and regulation of life processes, to maintain a relatively constant state of equilibrium, for example, blood composition, temperature, etc.
Hyperbaric oxygen therapy	is a therapeutic procedure often used for general and local tissue oxygenation against anaerobes, after gas poisoning, to secure the graft, frostbite, burns, radioactive damage, osteomyelitis, slow-healing wounds, sudden deafness, etc.
Hypercapnia	is a state of elevated partial pressure of the CO_2 in the blood above $pCO_2 > 45$ *mmHg*, referred here as the symptoms of CO_2 poisoning.
Hyperoxia	is a state of breathing respiratory medium, wherein the oxygen partial pressure exceeds 21 *kPa*, this term is used herein to cases of breathing with medium with the partial pressure of oxygen, which creates a toxic threat.

Item	Description
Hypocapnia	(also hypocarbia) is a state of reduced partial pressure of CO_2 in the blood below normal values.
Hypoxemia	is a reduction of the diffusion of the oxygen in the lungs.
Hypoxia	is the state of oxygen deficiency in the tissue resulting from reduced oxygen diffusion in the lungs (hypoxemia) or disorder of the blood oxygen transport to the tissues (ischemia).
Ischemia	is a disorder of oxygen transport by blood to the tissues.
Mathematical model	is a mathematical description of the behavior of the modelled system, usually expressed in the form of algebraic mathematical equations.
Metabolite	is an organic or inorganic metabolic product.
Mitochondria	is a structure surrounded by a membrane, present in the plasma of most nucleated cells, and is the place where in the process of cellular respiration most energy-carrying compounds are produced.
Model	is a system whose function is to imitate the distinguished features of another system, known as the original.
Mol	is a basic unit in the SI system that specifies the amount of substance in the system.[2]
Paresthesia	is a deficit in sensorial perception involving the wrong location of the stimulus and warped fillings described as tingling, numbness, etc.
Premix	is a mixture of two or more components standardized prior to use.[3]
Preoxygenation	is flushing the body with oxygen.[4]
Radical	is a group of atoms, generally incapable of independent existence, having unpaired electrons.[5]
Reaction kinetics	is reaction duration/mechanism as a function of time.
Reaction mechanism	describes a mechanism created during the chemical reaction via intermediate stages, which further result in final products, as summarized in reaction equation, illustrating transition from the starting particles into particles which are reaction products.
Respiratory dead space volume	is the lung space volume in which the ventilation does not occur or mass transfer occurs only to a limited extent.
Saturated solution	is a solution which, under specified conditions, is/can be in equilibrium with excess of the solute.
Semi-empirical methods	are methods describing the behavior of the object according to criterion equation[6] typically containing one or several constants determined empirically for a model, but these do not have constant physical interpretation and are often dependent on the units of measurement.[7]
Solution concentration	is the amount of solute in a given amount of solvent; concentration of the solution most often is expressed using a fraction or percentage concentration.[8]
Somatic	concerns the body; carnal, physical.
System	is distinguished from the reality collection of elements with links between them (Other elements are outside the system and form the system context. Links between system elements and context elements create interactions between the system and the environment).
Target strength	Strength is the most commonly expressed in decibels [$dB@1\ m$] as the ratio of the intensity of the wave [$W \cdot m^{-2}$] reflected from the target toward the receiver at a distance from its center and a intensity of the acoustic wave [$W \cdot m^{-2}$] incident on the object from the receiver.

NOTES

1 If the reaction reaches equilibrium it can be characterized by the equilibrium constant, which depends only on the type of reaction and temperature.
2 1 mole of the substance occurs when the number of particles is equal to the number of atoms contained in the carbon isotope ^{12}C with mass 0.012 kg.
3 This term is used here to emphasize that nitrox used in a diving apparatus must be premixed, seasoned, tested, certified and operationally tested before use.
4 In normal tissues dissolved nitrogen is in equilibrium with atmospheric air, but during the initial phase of diving nitrogen is purged from the tissues, causing delay in the dissolution of nitrogen in the tissues during the phase of nitrox diving, thus allowing a slight shortening of the decompression process without increasing the risk of decompression sickness (*DCS*).
5 Free valences.
6 Received, e.g. via dimensional analysis.
7 Algebraic model thus obtained is not adequate when you change the units of measurement, and due to the approximate nature of the relationship (consisting of approximately a function of the unknown laws of physics) semi-empirical and empirical models are valid only to the extent of their determination and allow in this field interpolation and every extrapolation is always risky.
8 There are several kinds of concentration expressed using a fraction or percentage concentration: mole (normal), mass, volume, etc.

Acronyms

Item	Description
CCR SCUBA	closed circuit rebreather self-contained breathing apparatus
CNSyn	oxygen toxicity impact on the central nervous system is called the Paul Bert effect; sometimes the acronym *CNS* for central nervous syndrome is used, but for the purposes of this book, in order to distinguish it from the acronym used to define the central nervous system (*CNS*), *CNSyn* will be used
CO_2	carbon dioxide
fsw	feet sea water
GABA	There are two *GABA* types of receptors binding γ-amino butyric acid: – $GABA_A$ adrenoceptor regulates the influx of chloride ions into the cell hindering the formation of action potentials responsible for providing information in the nervous system – $GABA_B$ adrenoceptor regulates the flow of potassium ions and calcium to neutralize the effect of chloride ions and regulates the release of neurotransmitters
HBO	hyperbaric medicine (*HBOT* – hyperbaric oxygen therapy) is the use of pressurized oxygen for medical purposes; in addition to the treatment of a decompression sickness pressure it is a good method for the treatment of gangrene, diabetic foot, exhaust gases and carbon monoxide poisoning and helps in the healing of wounds after frostbite and skin transplantation; there are also reports of its use in regenerating nerve tissue, especially the auditory nerve
IR	infrared
MoD	Ministry of Defense
OTT	oxygen tolerance test
pH	the exponent of multiplicity of hydrogen ions,[1] which is regarded as a physical quantity and a quantitative measure of the activity of hydrogen ions in solution
ppm	parts per million
SCR SCUBA	semiclosed circuit rebreather self-contained breathing apparatus
SEV	surface equivalent value
STANAG	NATO standardization agreement
STP	standard temperature and pressure – $T = 273\ K$ and $p = 101{,}325\ kPa$
TT	treatment table

NOTE

1 Negative logarithm of the hydrogen ion activity; pH of water and neutral solution $pH = 7.0$, acidic solutions $pH < 7$, alkaline $pH > 7$.

Introduction

This monograph contains the results of studies conducted within the framework of development project N° OR00000108 entitled "Designing Decompression in Combat Missions" and the development project N° OR00009811 entitled "Detection and Prevention of Diver Terrorist Threats." Both projects were financed by the Polish National Research Council. The monograph contains the results of the research on the possibilities of conducting nitrox[1] (Nx)/oxygen exposure using a diving apparatus type $oxy - CCR/Nx - SCR\ AMPHORA\ SCUBA$. It also contains references to the other published results dedicated to the design and the methods of testing semiclosed and closed circuit diving apparatuses. The reason for undertaking the research was the need to adapt the diving technology to the tasks planned for the future.

FRAMEWORK

Research and training was carried out under the authorization obtained from the Bioethics Committee of the Military Medical Institute. From the start, it was assumed that the technical capabilities of the apparatus should contribute to improving the flexibility of combat missions. The $oxy - CCR/Nx - SCR\ AMPHORA\ SCUBA$ apparatus has been designed to facilitate deployment of assault groups/special forces through the torpedo hatch of large submarines staying at periscope depth. The launch of an underwater mission at depths below 6 mH_2O with the use of oxygen apparatus type $oxy - CCR\ SCUBA$ significantly limits the duration time and the depth of the mission, which makes it less effective. On the other hand, resignation from the benefit of a hidden transfer and from the possibility to launch a mission of a special group/special section from the submarine is, from a tactical point of view, unreasonable exclusion. It was recognized that the best option would be to start the mission using Nx and semiclosed circuit $Nx - SCR\ SCUBA$ then proceed to a transit depth $H \le$ 6 mH_2O and finally continue the mission using oxygen in a closed circuit $oxy - CCR$ $SCUBA$. In general, the depth of the Nx^2 diving phase does not exceed 24 mH_2O^3 – see Figure 0.1.

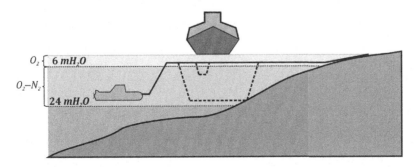

FIGURE 0.1 Typical dive profile for oxy/Nx dives

DOI: 10.1201/9781003309505-1

SUMMARY

The research program focused on two areas. The first was to introduce a method of transit-type dives using oxygen with the possibility of one trip below the depth of transit. The possibility of making a second trip of this kind, which would complete the dive, was also investigated. During this phase of research, the closed circuit mode of the *CCR* oxygen apparatus was used: *oxy – CCR AMPHORA SCUBA*.

In addition to evaluating the risks of oxygen toxicity that could occur during trips to a depth greater than the maximum depth of the transit, research was also focused on developing rules of purging the respiratory loop of the apparatus.

The second area covered the use of *oxy – CCR/Nx – SCR AMPHORA SCUBA* apparatus in *Nx* mode and was focused on developing tables and know-how for an *Nx* trip after the initial phase of transit. Throughout this initial phase, the apparatus is used in oxygen closed circuit mode: *oxy – CCR AMPHORA SCUBA*. During a dive to a depth greater than the maximum depth of the transit, the apparatus is switched to *Nx* supply in semiclosed circuit mode: *Nx – SCC AMPHORA SCUBA*. The main task of the research was to establish the maximum dive time while using *Nx*, assuming direct decompression obligation after reaching the specified maximum depth of the trip. While determining the maximum dive time, the influence of the initial preoxygenation[4] had to be taken into consideration.

NOTES

1 A nitrogen-oxygen gas mixture in which the oxygen content is different from that in the air.
2 Prior to the underwater mission which uses oxygen as a breathing medium.
3 For *Nx – SCR AMPHORA SCUBA* apparatus, *Nx* maximum operational depth can be deeper (see Chapter 1).
4 Flushing the body with oxygen. In standard conditions, nitrogen dissolved in tissues is in equilibrium with atmospheric air, but during the initial phase of oxygen diving it is flushed out from the tissues, causing delay in the dissolution of nitrogen in the tissues during the nitrox phase of diving, which allows a slight shortening of the process of decompression without the increased risk of decompression sickness.

1 Subject of the Research

Shown in Figure 1.1, alternatively semiclosed or closed circuit French diving apparatus *oxy – CCR/Nx – SCR AMPHORA SCUBA* is a successor of apparatus type *Oxy – Mix* (Figure 1.2).

An autonomous *SCUBA* apparatus type *AMPHORA* can operate in two configurations: a closed circuit mode with oxygen as a respiratory medium: *oxy – CCR AMPHORA SCUBA*, and in semiclosed circuit with *Nx* as breathing gas: *Nx – SCR AMPHORA SCUBA*. Modes can be changed by operating switch 15 during the dive (Figure 1.3).

1.1 BREATHING MEDIUM CIRCULATION

Proper breathing medium circulation is maintained by the unidirectional valves fitted in mouthpiece 9 in Figure 1.3. During exhalation, the exhaust valve of mouthpiece 9 opens and exhaled gases pass on through the mouthpiece, exhaust hose and CO_2 absorber 8 to breathing bag 6. By setting switch 15 for the oxygen supply,[1] the underpressure created during the inhalation phase forces the oxygen supply by second-stage regulator 10. Constant overpressure of oxygen in the second stage is provided by regulator 4, which senses the ambient pressure. The breathing medium supplemented with oxygen is sucked to the diver's lungs through an inhalation hose, the inhalation valve and mouthpiece 9. With closed oxygen circulation, relief valve 7[2] works only during the ascent phase, disposing the excess of expanding

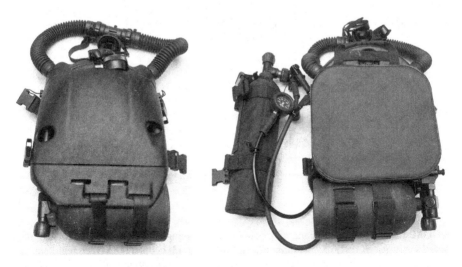

FIGURE 1.1 Alternatively semiclosed or closed circuit diving apparatus *oxy – CCR/Nx – SCR AMPHORA SCUBA*.

DOI: 10.1201/9781003309505-2

FIGURE 1.2   One of the latest versions of French diving apparatus type *Oxy – Mix*.

breathing medium with decreasing ambient pressure.[3] Bypass valve 12 is used for initial filling of the breathing system, refilling it during the dive and for purging the breathing loop of the apparatus.

If you set switch 15, controlling the type of breathing medium, opposite to what is indicated in Figure 1.3, the breathing space of apparatus is supplied with *Nx*.

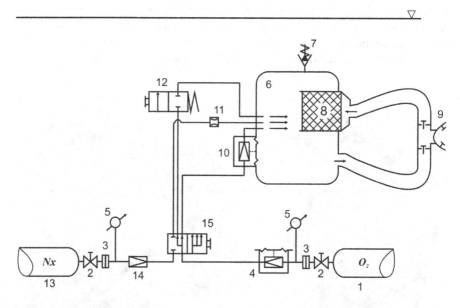

FIGURE 1.3   Principle of operation of a two-system diving apparatus *oxy – CCR/Nx – SCR AMPHORA SCUBA*: (1) oxygen tank, (2) shut-off valve, (3) couple, (4) oxygen regulator, (5) manometer, (6) breathing bag, (7) relief valve, (8) Co_2 absorber, (9) mouthpiece, (10) second-stage regulator, (11) orifice, (12) bypass valve, (13) *Nx* tank, (14) reducer for constant output pressure, (15) switching valve.

Exhaled, regenerated gas is mixed with fresh *Nx* dispensed through supply orifice 11, working at supercritical flow conditions, provided by the regulator with a constant value of the reduced pressure 14 (Kłos R., 2000; Kłos R., 2012; Kłos R., 2021). The amount of the gas supplied in a permanent-dispensing system is greater than the demand, hence relief valve 7 periodically[4] releases the excess of breathing medium into the water.

Partial recirculation of the regenerated breathing medium has an impact on its economical use, but it creates problems with the stabilization of the composition of the breathing medium inhaled by the diver. This generates problems with finding a safe and effective decompression procedure. There is a mathematical relationship between the inhaled breathing medium content and the depth range, the dosage of breathing medium and the diver's workload. When planning a dive,[5] the appropriate *premix*[6] and the dispensing orifice should be selected.[7] Once the choice is made, it is impossible to make changes during the dive, since the apparatus does not have an orifice switcher or additional cylinders with different types of premix.[8]

When switch 15 is selected for the premix supply, according to the manufacturer recommendations, second-stage regulator resistance 10 should be set to the maximum value, practically blocking its operation. Emerging breathing medium deficiency during descent should be supplemented by using bypass valve 12. The second-stage regulator does not have to be blocked, bearing in mind, however, that this could result in higher consumption of the breathing medium. This situation, in the absence of a cutoff device for reserve premix supply, creates an additional threat to the unwary diver.

1.2   VENTILATION PROCESS

In the premix-supplied apparatus with a semiclosed circulation *Nx – SCR AMPHORA SCUBA*, the relative falloff of the value of the oxygen content in the inhaled respiratory medium is noticed, as correlated to the oxygen content in the cylinders with premix. This is because partly used breathing medium mixes in the breathing space of the apparatus with a fresh supply of the premix (Figure 1.3).

Presented in Figure 1.4, the oxygen balance is proper for an apparatus with only one breathing bag. Oxygen with premix 1 is supplied to the breathing bag together with regenerated respiratory medium 2. Oxygen is spent with inhaled breathing medium 3 and through relief valve 4. Part of the oxygen from the inhaled breathing medium is consumed by diver's body 5 and part returns after regeneration to breathing bag 2.[9]

On the basis of Figure 1.4, the oxygen and breathing medium balance can be presented as shown in Table 1.1 (Kłos R., 2000; Kłos R., 2012; Kłos R., 2021).

Using the balance statement presented in Table 1.1, the dependence of oxygen content in the breathing bag as a function of time $x(O_2) = f(t)$ can be calculated, assuming that the values of the gas dosage $\dot{V}$, oxygen consumption $\dot{\upsilon}$ and lung ventilation $\dot{V}_E$ do not depend on the depth of dive: $\dot{V} \neq f(H)$, $\dot{\upsilon} \neq f(H)$ and $\dot{V}_E \neq f(H)$, where H represents the depth of dive.[10] It should be noted that $\dot{V}$, $\dot{\upsilon}$ and $\dot{V}_E$ are the values related to standard pressure[11] p_0, and therefore the number of moles of oxygen

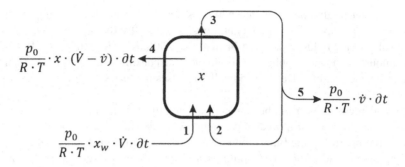

FIGURE 1.4 The balance of oxygen mole in the breathing space of $Nx - SCR\ AMPHORA$ $SCUBA$, where: $\dot{V}$ – fresh breathing medium dosing, $\dot{\upsilon}$ – a stream of consumed oxygen, $\dot{V}_E$ – lung ventilation, V – breathing bag volume, x_w – oxygen mole fraction in premix, dt – basic period of time, i – next period of time dt, for which the balance is run, $x(i)$ – oxygen mole fraction in the breathing bag for moment $t = i \cdot dt$, $n(i)$ – number of oxygen moles in the breathing bag for moment $t = i \cdot dt$, p_0 – normal pressure, R – universal gas constant, T – temperature in absolute temperature scale.

TABLE 1.1
Oxygen and breathing medium molar balance in the breathing bag – symbols used as in Figure 1.4 (Kłos R., 2021).

		With the fresh breathing medium	With breathing circulation	Through the relief valve
Oxygen	Increase	$\dfrac{p_0}{R \cdot T} \cdot \dot{V} \cdot x_w \cdot \partial t$	$\dfrac{p}{R \cdot T}\left[\dot{V}_E \cdot x(i) - \dfrac{p_0}{p} \cdot \dot{v}\right]\partial t$	–
	Decrease	–	$\dfrac{p}{R \cdot T} \cdot \dot{V}_E \cdot x(i) \cdot \partial t$	$\dfrac{p_0}{R \cdot T} \cdot \left(\dot{V} - \dot{v}\right) \cdot x(i) \cdot \partial t$
The whole breathing medium	Increase	$\dfrac{p_0}{R \cdot T} \cdot \dot{V} \cdot \partial t$	$\dfrac{p}{R \cdot T} \dot{V}_E - \dfrac{p_0}{R \cdot T} \cdot \dot{v} \cdot \partial t$	–
	Decrease	–	$\dfrac{p_0}{R \cdot T} \cdot \dot{V}_E \cdot \partial t$	$\dfrac{p_0}{R \cdot T} \cdot \left(\dot{V} - \dot{v}\right) \cdot x(i) \cdot \partial t$

Where: $\dot{V}$ – metering of fresh breathing medium, $\dot{v}$ – oxygen consumption, $\dot{V}_E$ – pulmonary ventilation, x_w – oxygen molar fraction in the *premix*, ∂t – elementary time, i – successive time interval ∂t for substance balance, $x(i)$ – oxygen molar fraction in the breathing bag at moment $t = i \cdot \partial t$, p_0 normal pressure, R – universal gas constant, T – absolute temperature.

associated with these values should be related to the *STP* conditions. As indicated in Figure 1.4 and Table 1.1, the dependence of the mole fraction of oxygen $x(O_2)$ in one bag of the breathing apparatus with semiclosed breathing medium circulation, as a function of time $x(O_2) = f(t)$, is derived with the method of convergent series calculations in the form of mathematical proof shown in Table 1.2 (Kłos R., 2021). The same result can be obtained using the integral method – Table 1.3 (Kłos R., 2021).

TABLE 1.2
The relationship between oxygen molar fraction *x* and time *t* for the single-bag semiclosed circuit diving apparatus *SCR* with constant metering of premix by means of limits calculus based on oxygen molar balance in the breathing bag presented in Table 1.1 (Kłos R., 2021).

A: 1° $\dot{V} \neq f(H)$; $\dot{\upsilon} \neq f(H)$; $\dot{V}_E \neq f(H)$

2° $\dot{V} = \dot{V}_0 = \dot{V}(p = p_0)$; $\dot{\upsilon} = \dot{\upsilon}_0 = \dot{\upsilon}(p = p_0)$

T: $x(O_2) = f(t)$

P: 1° For elementary time $d\tau$ at time $(i + 1)\, d\tau$ it can be written: *from oxygen balance*

$$n(i+1) = (n(i) + \frac{p_0}{R \cdot T} \cdot \left[x_w \cdot \dot{V} + x(i) \cdot \dot{V} - \dot{\upsilon} + x(i) \cdot \dot{\upsilon} \right] \partial t$$

2° $x(i+1) = \left[x(i) - \frac{x_w \cdot \dot{V} - \dot{\upsilon}}{\dot{V} - \dot{\upsilon}} \right] \left(1 - \frac{p_0}{p} \cdot \frac{\dot{V} - \dot{\upsilon}}{V} \cdot \partial t \right) + \frac{x_w \cdot \dot{V} - \dot{\upsilon}}{\dot{V} - \dot{\upsilon}}$ *from 1° divided by:*

$$n = \frac{p \cdot V}{R \cdot T} \text{ and}$$
$$x(i) = \frac{n(i)}{n}$$

3° $x(i+2) = \left[x(i+1) - \frac{x_w \cdot \dot{V} - \dot{\upsilon}}{\dot{V} - \dot{\upsilon}} \right] \left(1 - \frac{p_0}{p} \cdot \frac{\dot{V} - \dot{\upsilon}}{V} \cdot \partial t \right) + \frac{x_w \cdot \dot{V} - \dot{\upsilon}}{\dot{V} - \dot{\upsilon}}$ *from 2°*

4° $x(i+2) = \left[x(i) - \frac{x_w \cdot \dot{V} - \dot{\upsilon}}{\dot{V} - \dot{\upsilon}} \right] \left(1 - \frac{p_0}{p} \cdot \frac{\dot{V} - \dot{\upsilon}}{V} \cdot \partial t \right)^2 + \frac{x_w \cdot \dot{V} - \dot{\upsilon}}{\dot{V} - \dot{\upsilon}}$ *from 3° and 2°*

5° $x(i+j) = \left[x(i) - \frac{x_w \cdot \dot{V} - \dot{\upsilon}}{\dot{V} - \dot{\upsilon}} \right] \left(1 - \frac{p_0}{p} \cdot \frac{\dot{V} - \dot{\upsilon}}{V} \cdot \partial t \right)^j + \frac{x_w \cdot \dot{V} - \dot{\upsilon}}{\dot{V} - \dot{\upsilon}}$ *form 2° ÷ 4°*

6° For $i = 0$, $j = t$ and $x(0) = x_0$ it can be written:

$$x(t) = \frac{x_w \cdot \dot{V} - \dot{\upsilon}}{\dot{V} - \dot{\upsilon}} + \left[x_0 - \frac{x_w \cdot \dot{V} - \dot{\upsilon}}{\dot{V} - \dot{\upsilon}} \right] \cdot \lim_{j \to \infty} \left(1 - \frac{p_0}{p} \cdot \frac{\dot{V} - \dot{\upsilon}}{V} \cdot \partial t \right)^j$$ *from 5°*

from 6°

7° $x(t) = a + b \cdot \lim_{j \to \infty} \left[\left(1 + \frac{c}{j} \right)^{j/c} \right]^c$ $\lim_{j \to \infty} (j \cdot dt) = t$

where: $a = \frac{x_w \cdot \dot{V} - \dot{\upsilon}}{\dot{V} - \dot{\upsilon}}$, $b = x_0 - a$, $c = -\frac{p_0}{p} \cdot \frac{\dot{V} - \dot{\upsilon}}{V} \cdot j \cdot \partial t$

8° $x(t) = exp(c) + a$, because: *from 7°*

$$\forall_{a(n) \neq 0} \lim_{n \to \infty} a(n) = 0 \Rightarrow \lim_{n \to \infty} [(1 + a(n)]^{1/a(n)} \equiv e$$ $a(n) = \frac{c}{j}$

from 7° ÷ 8°

9° $x(t) = \frac{x_w \cdot \dot{V} - \dot{\upsilon}}{\dot{V} - \dot{\upsilon}} + \left(x_0 - \frac{x_w \cdot \dot{V} - \dot{\upsilon}}{\dot{V} - \dot{\upsilon}} \right) \cdot exp\left(-\frac{p_0}{p} \cdot \frac{\dot{V} - \dot{\upsilon}}{V} \cdot t \right)$ *q.e.d.*

A – assumption, T – thesis, P – proof (evidence)

Where: $\dot{V}$ – metering of fresh breathing medium, $\dot{\upsilon}$ – oxygen consumption, $\dot{V}_E$ – pulmonary ventilation, x_w – oxygen molar fraction in the premix, x_0 – initial composition of breathing medium in breathing bag ∂t – elementary time, i – successive time interval ∂t for substance balance, $x(i)$ oxygen molar fraction in the breathing bag moment $t = i \cdot \partial t$, $n(i)$ – number of oxygen moles in the breathing bag at moment $t = i$ ∂t, p_0 – normal pressure, p – pressure at depth of diving R – universal gas constant, T – the absolute temperature, V – breathing bag volume.

TABLE 1.3
Derivation of molar oxygen fraction $x(O_2)$ in function of time $f(t)$ for a single-bag diving apparatus with semiclosed circulation and constant dosing of the breathing agent by means of an integral calculus as in Figure 1.4 (Kłos R., 2021).

A: 1° $\dot{V} \neq f(H); \dot{\upsilon} \neq f(H); \dot{V}_E \neq f(H)$

      2° $\dot{V} = \dot{V}_0 = \dot{V}(p = p_0); \dot{\upsilon} = \dot{\upsilon}_0 = \dot{\upsilon}(p = p_0)$

T: $x(O_2) = f(t)$

P: 1° $\dfrac{p}{R \cdot T} \cdot V \cdot \dfrac{\partial x}{\partial t} = \dfrac{p_0}{R \cdot T} \cdot \dot{V} \cdot x_w - \dfrac{p_0}{R \cdot T} \cdot \dot{\upsilon} - \dfrac{p_0}{R \cdot T} \cdot (\dot{V} - \dot{\upsilon}) \cdot x$ from oxygen balance breathing bag

      2° $\underset{a \equiv \frac{p_0}{p} \cdot \frac{\dot{V} \cdot x_w - \dot{\upsilon}}{V} = idem; \, b \equiv \frac{p_0}{p} \cdot \frac{\dot{V} - \dot{\upsilon}}{V} = idem}{\forall} \dfrac{\partial x}{\partial t}$ as for the stable conditions: $\dot{V}, \dot{V}_E, \dot{\upsilon} = const$

              $= \dfrac{p_0}{p} \cdot \dfrac{\dot{V} \cdot x_w - \dot{\upsilon}}{V} - \dfrac{p_0}{p} \cdot \dfrac{\dot{V} - \dot{\upsilon}}{V} \cdot x$

      3° $\underset{a \equiv \frac{p_0}{p} \cdot \frac{\dot{V} \cdot x_w - \dot{\upsilon}}{V} = idem; \, b \equiv \frac{p_0}{p} \cdot \frac{\dot{V} - \dot{\upsilon}}{V} = idem}{\forall} \partial x = (a - b \cdot x) \partial t$ from 2°

      4° $\dfrac{\partial x}{b \cdot x - a} + \partial t = 0$ from 3° by dividing by $(b \cdot x - a)$

      5° $\int \dfrac{\partial x}{b \cdot x - a} + \int \partial t = C$ from 4° and integral definition

          where: C– integral constant

      6° $\dfrac{1}{b} \cdot \ln|b \cdot x - a| + t = C; \ln|b \cdot x - a| \equiv b \cdot (C - t) = C' - b \cdot t$ From 5°

          where– C'– the new constant

      7° $\exp(C' - b \cdot t) = C'' \cdot \exp(-b \cdot t) \equiv b \cdot x - a$ where–C'-the new constant   from 6° and natural logarithm definition

      8°   If for boundary conditions $t \to 0 \Rightarrow x \to x_0$, then: From 7°
          $C'' = b \cdot x_0 - a \Rightarrow (b \cdot x_0 - a) \cdot \exp(-b \cdot t) = b \cdot x - a$

      9° $x(t) = \dfrac{x_w \cdot \dot{V} - \dot{\upsilon}}{\dot{V} - \dot{\upsilon}} + \left(x_0 - \dfrac{x_w \cdot \dot{V} - \dot{\upsilon}}{\dot{V} - \dot{\upsilon}} \right) \cdot \exp\left(-\dfrac{p_0}{p} \cdot \dfrac{\dot{V} - \dot{\upsilon}}{V} \cdot t \right)$ from 2° to 8° q.e.d.

A – assumption, T – thesis, P – proof (evidence)

Where: $\dot{V}$ – metering of fresh breathing medium, $\dot{\upsilon}$ – oxygen consumption, x_w – oxygen molar fraction in premix, $x(t)$ – oxygen molar fraction in breathing gas mix by diver at moment t, x_0 – the initial composition of breathing medium in breathing bag, ∂t – elementary time, p_0 – normal pressure, p – pressure at depth of diving, R – universal gas constant, T – absolute temperature, V – breathing bag volume.

The analysis of equation (9°) presented in Table 1.3 leads to two conclusions: $x(t = 0) = x_0 \wedge \lim_{t \to \infty} x(t) = \frac{x_w \cdot \dot{V} - \dot{\upsilon}}{\dot{V} - \dot{\upsilon}}$, where the condition $x(t = 0) = x_0$ is a simple consequence of the assumptions (6°) presented in Table 1.3, while relationship $\lim_{t \to \infty} x(t) = \frac{x_w \cdot \dot{V} - \dot{\upsilon}}{\dot{V} - \dot{\upsilon}}$ is quoted in the literature with regard to a dynamic gas equilibrium model resulting from the mass balance of oxygen and the breathing medium in the respiratory bag, assuming perfect mixing conditions (Williams S., 1975; Haux G., 1982).

1.3 DESIGN

Traditional semiclosed circuit diving apparatuses with constant dosage of breathing medium were designed on the basis of the model derived from the global mass balance (Williams S., 1975; Haux G., 1982): $x_s = \frac{x_w \cdot \dot{V} - \dot{\upsilon}}{\dot{V} - \dot{\upsilon}}$, where x_s is stable oxygen content in the inhaled breathing medium [$mol \cdot mol^{-1}$]. The stable value of oxygen mole fraction x_s with the values of dosed flow rate of premix $\dot{V} > 5\ dm^3 \cdot min^{-1}$ is reached within a few minutes (Kłos R., 2000; Kłos R., 2021). Using the definition of partial pressure, the stable value of oxygen partial pressure p_s in the inhaled breathing medium can be expressed by the formula:

$$\forall_{\upsilon=idem}\ p_s = x_s \cdot p = \frac{x_w \cdot \dot{V} - \dot{\upsilon}}{\dot{V} - \dot{\upsilon}} \cdot p \tag{1.1}$$

where: p_s – partial oxygen pressure in inhaled breathing medium, p – absolute pressure at depth of dive, x_s – stable value of oxygen contents in breathing mixture, x_w – oxygen contents in premix, $\dot{V}$ – dosed flow rate of premix, $\dot{\upsilon}$ – oxygen consumption.

From equation (1.1), one can calculate dosage values $\dot{V}$ and the mole fraction of oxygen in premix x_w, with the assumed boundary values of pressure at dive depth p, value of consumption $\dot{\upsilon}$ and stable partial pressure of oxygen p_s. The maximum value of oxygen partial pressure p_s^{max} is achieved for the minimum of consumption $\dot{\upsilon}^{min}$ and maximum absolute pressure p^{max} at the depth of dive.[12] The minimum value of oxygen partial pressure p_s^{min} is achieved at the maximum of its consumption $\dot{\upsilon}^{max}$ and minimum pressure p^{min} at the depth of dive.[13] From here, you can write the following system of equations:

$$\begin{cases} p_s^{max} = \dfrac{x_w \cdot \dot{V} - \dot{\upsilon}^{min}}{\dot{V} - \dot{\upsilon}^{min}} \cdot p^{max} \\[4mm] p_s^{min} = \dfrac{x_w \cdot \dot{V} - \dot{\upsilon}^{max}}{\dot{V} - \dot{\upsilon}^{max}} \cdot p^{min} \end{cases} \tag{1.2}$$

where: p_s^{max} – maximum partial pressure of oxygen in inhaled breathing medium, p_s^{min} – minimum partial pressure of oxygen in inhaled breathing medium, p^{max} – absolute pressure at maximum diving depth, p^{min} – absolute pressure at minimum diving depth, $\dot{v}^{min}$ – minimum of oxygen consumption, $\dot{v}^{max}$ – maximum oxygen consumption.

In the system of equations (1.2), pressure at diving depths $\{p^{min}; p^{max}\}$, stable oxygen partial pressures $\left\{p_s^{min}; p_s^{max}\right\}$ and oxygen consumption rates $\left\{\dot{v}^{min}; \dot{v}^{max}\right\}$ are specified using the proposed tables and know-how. The range of pressure at the depths of work results from the allowed range of diving depths. The values of oxygen consumption rates have been determined experimentally for the defined type of exercise and can be found in literature[14] (Przylipiak M., Torbus J., 1981; Kłos R., 2021). The allowed range of the oxygen partial pressure based on medical research is presented in the next chapters. The values determined from the system of equations (1.2) are premix dosage rate $\dot{V}$ and oxygen concentration in premix x_w.

To obtain a carefully composed breathing medium is a cumbersome and expensive operation. From a practical point of view, introducing limits to the permitted dosages and oxygen concentrations instead of recommending one specific value is convenient. Taking into account these parameters requires extending relation (1.2) by permissible deviations from the nominal values of dosage by $\Delta\dot{V}$ and oxygen content by Δx_w in the premix. Taking into account the permissible deviations from these nominal values leads to a system of equations:

$$\begin{cases} p_s^{max} = \dfrac{\left(x_w + \Delta x_w\right) \cdot \left(\dot{V} + \Delta\dot{V}\right) - \dot{v}^{min}}{\dot{V} + \Delta\dot{V} - \dot{v}^{\dot{v}^{max}}} \cdot p^{max} \\[3em] p_s^{min} = \dfrac{\left(x_w + \Delta x_w\right) \cdot \left(\dot{V} + \Delta\dot{V}\right) - \dot{v}^{max}}{\dot{V} + \Delta\dot{V} - \dot{v}^{\dot{v}^{max}}} \cdot p^{min} \end{cases} \qquad (1.3)$$

where: Δx_w – permissible deviation from accepted as the optimum mole fraction of oxygen in the premix, $\Delta\dot{V}$ – permissible deviation from accepted as the optimal values of the flow rate of dosed premix.

Solving the system of equations (1.3), by eliminating the oxygen content from the premix x_w a quadratic equation is obtained with the dosage $\dot{V}$. Of all solutions of the system of equations (1.3), only those which have physical sense (Figure 1.5) should be chosen.

The diving apparatus with a semiclosed circuit and with constant dosage of premix is designed mostly for several premixes and composition dosages related to them, because there is no one universal solution for a sufficiently wide range of safe diving depths.[15] In other words, it has not been possible to design an effective and safe diving apparatus for a wide range of diving depths, fed by only one type of premix[16] (Table 1.4).

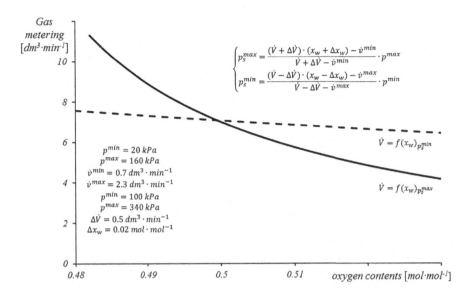

FIGURE 1.5 Example of calculation of dosage $\dot{V}$ and oxygen content x_w for the diving semiclosed circuit apparatus with constant dosing of premix Nx – *SCR AMPHORA SCUBA* according to (1.3), where: Δx_w – permissible deviation from accepted as the optimum value of oxygen mole fraction in the premix, $\Delta\dot{V}$ –permissible deviation from accepted as the optimum value of the flow of dosed premix.

TABLE 1.4
Optimal $\dot{V}$ dosage and the mole fractions of oxygen x_w in the premix.

Minimum oxygen partial pressure						$= 20\ kPa$						
Maximum oxygen partial pressure						$= 160\ kPa$						
Minimum minute oxygen usage						$= 0.7\ dm^3 \cdot min^{-1}$						
Maximum minute oxygen usage						$= 2.8\ dm^3 \cdot min^{-1}$						
Dosage tolerance						$= \pm 2.0\ dm^3 \cdot min^{-1}$						
Oxygen contents tolerance						$= \pm 0.02\ mol \cdot mol^{-1}$						

H_{max} [mH_2O]	H_{min} [mH_2O] 0	10	20	30	40	50	60	70	80	90	100	110
150 x_w				0.099	0.101	0.103	0.104	0.105	0.105	0.106	0.106	0.106
$\dot{V}$				70.82	54.56	47.59	43.71	41.24	39.53	38.28	37.32	36.56
140 x_w				0.107	0.109	0.111	0.112	0.113	0.113	0.114	0.114	0.114
$\dot{V}$				58.62	47.29	42.09	39.10	37.17	35.81	34.80	34.02	33.41
130 x_w			0.112	0.116	0.119	0.120	0.121	0.122	0.122	0.123	0.123	0.123
$\dot{V}$			75.33	49.16	41.15	37.27	34.98	33.46	32.39	31.59	30.96	30.47
120 x_w			0.123	0.127	0.129	0.131	0.132	0.132	0.133	0.133	0.133	0.134
$\dot{V}$			58.04	41.61	35.91	33.01	31.26	30.08	29.24	28.61	28.11	27.72

(Continued)

TABLE 1.4 *(Continued)*

Optimal $\dot{V}$ dosage and the mole fractions of oxygen x_w in the premix.

110	x_w		0.136	0.140	0.142	0.143	0.144	0.144	0.145	0.145	0.145
	$\dot{V}$		46.09	35.44	31.37	29.22	27.89	26.99	26.33	25.83	25.44
100	x_w	0.143	0.151	0.154	0.156	0.157	0.158	0.158	0.159	0.159	
	$\dot{V}$	78.67	37.31	30.31	27.41	25.82	24.82	24.13	23.63	23.25	
90	x_w	0.161	0.168	0.171	0.173	0.174	0.174	0.175	0.175		
	$\dot{V}$	51.40	30.60	25.96	23.91	22.75	22.01	21.50	21.12		
80	x_w	0.183	0.189	0.192	0.193	0.194	0.195	0.195			
	$\dot{V}$	36.68	25.29	22.22	20.79	19.97	19.43	19.05			
70	x_w	0.209	0.214	0.217	0.218	0.219	0.219				
	$\dot{V}$	27.45	20.98	18.98	18.00	17.42	17.04				
60	x_w	0.242	0.246	0.248	0.249	0.250					
	$\dot{V}$	21.10	17.40	16.12	15.48	15.09					
50	x_w	0.269	0.285	0.288	0.290	0.290					
	$\dot{V}$	40.05	16.44	14.36	13.58	13.18					
40	x_w	0.331	0.342	0.345	0.346						
	$\dot{V}$	20.49	12.86	11.75	11.31						
30	x_w	0.418	0.425	0.427							
	$\dot{V}$	12.74	10.00	9.47							
20	x_w	0.554	0.557								
	$\dot{V}$	8.50	7.63								
10	x_w	0.840									
	$\dot{V}$	3.50									

x_w – fraction of oxygen content $[m^3 \cdot m^{-3}]$

$\dot{V}$ – dosage $[dm^3 \cdot min^{-1}]$

The oxygen content in premix x_w and related to Nx choice of the rate of dosages $\dot{V}$ recommended by the manufacturer of the apparatus Nx – *SCR AMPHORA SCUBA* are given in Table 1.5 (AQUA LUNG, 2004).

During the research leading to expertise for diving with the use of oxygen as the main breathing medium with possibility of Nx trip, one type of premix Nx 0.43 with the dosage at the level $\dot{V} = 10.5 \, dm^3 \cdot min^{-1}$ was used (Table 1.4). Making oxygen experimental dives with an Nx trip for the depth range $H \in [6, 32] \, mH_2O$ are allowed.

1.4 STABILIZATION

Stabilization and homogenization of the breathing medium composition is a derivative of the respiratory ventilation and is important for safety, efficiency and economy of the process of diving. It also makes an important impact on the safety of decompression. These issues also play an important role in the design of the apparatus. An example of theoretical oxygen content $x(t)$ in the medium inhaled by the diver from the one bag of a semiclosed version of the apparatus as function of time t and rate of oxygen consumption $\dot{\nu}$ while breathing at surface $p = p_0$ is shown in Figure 1.6.[17]

TABLE 1.5
Nitrox contents and their association with the choice of dosage for apparatus Nx – SCR AMPHORA SCUBA (AQUA LUNG, 2004).

Nitrox	Dosage	H_{max}	Theoretical safety time[†]
	±0.5 dm³ · min⁻¹		
	[$dm^3 \cdot min^{-1}$]	[mH_2O]	[min]
$60\%_v O_2/N_2$	5.5	18	66(50)
$50\%_v O_2/N_2$	7.2	24	51(38)
$40\%_v O_2/N_2$	10.5	32	36(27)

† calculated for tank volume $V = 2.0\ dm^3$ and working pressure $p = 20\ MPa$; in brackets theoretical safety time after deducting 5 MPa of emergency pressure supply of reserve Nx

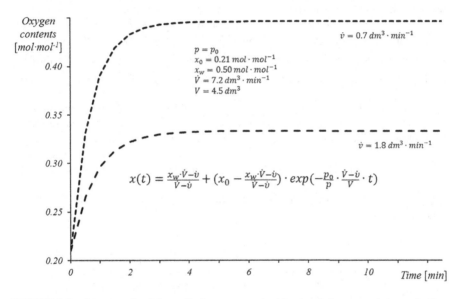

$$x(t) = \frac{x_w \cdot \dot{V} - \dot{v}}{\dot{V} - \dot{v}} + \left(x_0 - \frac{x_w \cdot \dot{V} - \dot{v}}{\dot{V} - \dot{v}}\right) \cdot \exp\left(-\frac{p_0}{p} \cdot \frac{\dot{V} - \dot{v}}{V} \cdot t\right)$$

FIGURE 1.6 An example of theoretical oxygen content in circulating gas mixture x in time function $x = f(t)$ and stream of oxygen consumed during respiration $\dot{v}$ on surface for apparatus Nx – $SCR\ AMPHORA\ SCUBA$, where: $\dot{V}$ – dosage of premix [$dm^3 \cdot min^{-1}$], $\dot{v}$ – stream of oxygen consumed [$dm^3 \cdot min^{-1}$], V – breathing bag volume [dm^3], x_0 – initial oxygen content of the breathing bag expressed as molar fraction [$mol \cdot mol^{-1}$], x_w – oxygen mole fraction in premix [$mol \cdot mol^{-1}$], x – molar fraction of oxygen in the breathing bag for time t [$mol \cdot mol^{-1}$], p_0 – normal pressure [kPa].

Using the model of ventilation for breathing loop of Nx – $SCR\ AMPHORA$ $SCUBA$: $x(t) = \frac{x_w \cdot \dot{V} - \dot{v}}{\dot{V} - \dot{v}} + \left(x_0 - \frac{x_w \cdot \dot{V} - \dot{v}}{\dot{V} - \dot{v}}\right) \cdot \exp\left(-\frac{p_0}{p} \cdot \frac{\dot{V} - \dot{v}}{V} \cdot t\right)$ two criteria can be constructed for calculating the time required to stabilize the breathing atmosphere composition.

The first is the absolute stabilization criteria, which can be written as $\Delta x =$ $|x(t) - x(\infty)|$, where Δx represents accepted highest deviation of oxygen content in the inhaled breathing medium $x(t)$ from the asymptotic value $x(\infty)$. Using the model of ventilation, this criterion can be expressed by the relationship as follows: $\Delta x = \left| \frac{x_0 \cdot \dot{V} - x_0 \cdot \dot{\upsilon} - x_w \cdot \dot{V} + \dot{\upsilon}}{\dot{V} - \dot{\upsilon}} \right| \cdot \exp\left(-\frac{p_0}{p} \cdot \frac{\dot{V} - \dot{\upsilon}}{V} \cdot \Delta t_s\right)$. Converting it to get the time necessary

to stabilize the composition of the breathing medium Δt_s, the relationship can be rewritten as:

$$\Delta t_s \left(\Delta x\right) = \frac{p \cdot V}{p_0 \cdot \left(\dot{V} - \dot{\upsilon}\right)} \cdot \ln \frac{\left|x_0 - x_w\right| \cdot \dot{V} + \left(1 + x_0\right) \cdot \dot{\upsilon}}{\Delta x \cdot \left(\dot{V} - \dot{\upsilon}\right)} \tag{1.4}$$

where: Δx – accepted highest deviation of oxygen content in inhaled breathing medium $x(t)$ from asymptotic value $x(\infty)$, x_0 – initial oxygen content in the breathing bag expressed in mole fraction, x_w – oxygen mole fraction in premix, Δt_s – time required to stabilize the composition of the respiratory medium.

The absolute stabilization criterion is not convenient for comparing the time of stabilization of breathing medium compositions. It is better for this purpose to use the relative value, which can be written as $\delta_x = \frac{|x(t) - x(\infty)|}{x(\infty)} \cdot 100\%$, where:

δx – relative value of limit of deviation of oxygen content in breathing loop $x(t)$ from asymptotic value $x(\infty)$. As in the previous calculations, using our model of ventilation in the breathing loop of apparatus, the following can be written:

$\frac{\delta_x}{100\%} = \left[\frac{x_0 \cdot (\dot{V} - \dot{\upsilon})}{x_w \cdot \dot{V} - \dot{\upsilon}} - 1\right] \cdot \exp\left(-\frac{p_0}{p} \cdot \frac{\dot{V} - \dot{\upsilon}}{V} \cdot \Delta t_s\right)$. Extracting the time needed to stabilize the

composition of breathing medium Δt_s, the final dependence takes the form of equation as follows:

$$\Delta t_s \left(\delta_x\right) = \frac{p \cdot V}{p_0 \cdot \left(\dot{V} - \dot{\upsilon}\right)} \cdot \ln \frac{100 \cdot \left[\left|x_0 - x_w\right| \cdot \dot{V} + \left(1 + x_0\right) \cdot \dot{\upsilon}\right]}{\delta_x \cdot \left(x_w \cdot \dot{V} - \dot{\upsilon}\right)} \tag{1.5}$$

Analyzing the relationships (1.4)–(1.5), a conclusion can be made that the stabilization time is proportional to pressure p under which the diver operates. Time Δt_s required to stabilize the composition of the breathing medium in the breathing loop at a pressure greater than that of atmospheric is associated with the stabilization time at the atmospheric pressure as indicated by the following equation: $\Delta t_s (p > p_0) = \frac{p}{p_0} \cdot \Delta t_s \left(p = p_0\right)$. This equation shows that the stabilization of breathing medium at a lower pressure is quicker than at a higher pressure (Figure 1.7).

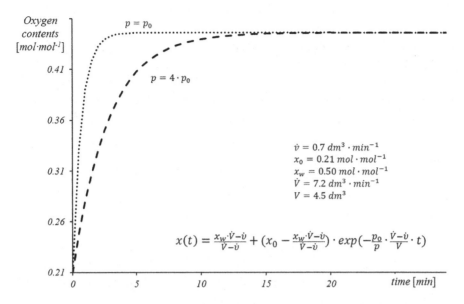

FIGURE 1.7   An example of theoretical process of stabilization of oxygen content x in respiratory medium as function of time t and rate of oxygen consumption $\dot{v}$ during respiration on surface $p = p_0$ and at depth of 30 $mH_2O \Rightarrow p = 4 \cdot p_0$ for $Nx - SCR\ AMPHORA\ SCUBA$.

1.5   RINSING

The oxygen content in the inhaled breathing medium for $Nx - SCR\ AMPHORA$
$SCUBA$ can be increased by purging the breathing loop with premix by using bypass
valve 12 (Figure 1.3). The purging procedure might be as follows:

- the diver possibly sucks the entire content of the breathing bag from the
  breathing loop through the mouthpiece and then exhales it[18]
- with the help of the bypass valve, the entire breathing loop is filled with
  fresh breathing medium
- if purging is applied several times, the previous procedure is repeated

In the course of purging,[19] the diver is holding the mouthpiece in the mouth during
the whole procedure or he/she can close the mouthpiece before filling in the breathing
loop. It should be noted that stabilization of the breathing medium composition in the
breathing loop depends on the dynamic process of mixing the regenerated breathing
medium with the fresh premix and is a function of the rate of oxygen consumption.

Purging of the breathing loop with premix may not ensure immediate occurrence
of the stabilization condition. This will, however, make the oxygen content in the
breathing loop comparable to the stabilized value achieved when the diver is not
working.

TABLE 1.6
The molar balance of oxygen and respiratory medium in the process of purging the breathing loop.

		Purging	Filling
Oxygen	Increases	–	$\frac{p}{p_0}\cdot\left(V_c-V_z\right)\cdot x_w$
	Decreases	$\frac{p}{p_0}\cdot\left(V_c-V_z\right)\cdot x_i$	–
	Remains	$\frac{p}{p_0}\cdot V_z\cdot x_i$	$\frac{p}{p_0}\cdot V_z\cdot x_i+\left(V_c-V_z\right)\cdot x_w$
Breathing medium		–	$\frac{p}{p_0}\cdot\left(V_c-V_z\right)$
	Increases	$\frac{p}{p_0}\cdot\left(V_c-V_z\right)$	–
	Decreases	$\frac{p}{p_0}\cdot V_z$	$\frac{p}{p_0}\cdot V_c$

Where: V_c – total volume of breathing loop, V_z – total volume of dead space in breathing loop, x_0 – initial oxygen content in breathing loop expressed in mole fraction, x_w – oxygen mole fraction in premix, j – multiplicity of purging.

The basis for deriving the equation of oxygen content x in multipurging conditions j[20] is the molar balance of oxygen and the balance of the whole volume of breathing medium. This balance is shown in Table 1.6.

According to this balance, one can write that for a single purge $j = i + 1$, the oxygen content of the breathing loop will be $x_{i+1} = \dfrac{\frac{p}{p_0}\left[V_z\cdot x_i+\left(V_c-V_z\right)\cdot x_w\right]}{\frac{p}{p_0}\cdot V} = \dfrac{V_z}{V_c}\cdot\left(x_i-x_w\right)+x_w$,

where V_c – total volume of the breathing loop, V_z – total volume of the dead space.[21] Similarly, the oxygen content for x_{i+2} in the breathing loop after $j = i + 2$ purging will be $x_{i+2} = \left(\dfrac{V_z}{V_c}\right)^2\cdot\left(x_i-x_w\right)+x_w$. Comparing the equation for $j = i + 1$ to the equation for $j = i + 2$, one can write: $x_{i+j} = \left(\dfrac{V_z}{V_c}\right)^j\cdot\left(x_i-x_w\right)+x_w$. For $i = 0$, this relationship can be simplified to the form:

$$x_j = x_w+\left(x_0-x_w\right)\cdot\left(\frac{V_z}{V_c}\right)^j \qquad (1.6)$$

where: V_c – total volume of breathing loop, V_z – total volume of dead space in breathing loop, x_0 – initial oxygen content in breathing loop expressed in mole fraction, x_w – oxygen mole fraction in premix, j – multiplicity of purging.

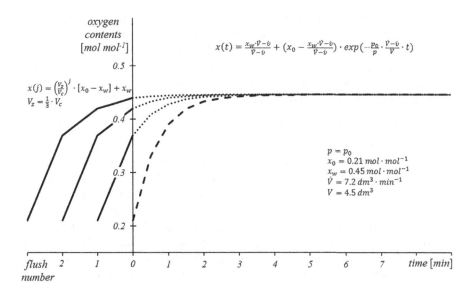

FIGURE 1.8 The theoretical composition of the breathing medium in the breathing loop of the $Nx - SCR\ AMPHORA\ SCUBA$ on surface as a function of time, after purging and without purging.

From equation (1.6) an important conclusion is drawn, namely that oxygen content x_j after j – number of purging acts is independent of depth $x_j \neq f(H)$, therefore purging of the breathing loop is equally effective at any depth. The theoretical content of oxygen in the breathing medium versus time for $Nx - SCR\ AMPHORA\ SCUBA$, after work without the initial purging act, is shown in Figure 1.8. In the calculation, it was assumed that 25% of the breathing loop is not exchanged for the premix. Figure 1.8 indicates that the relative efficiency of purging decreases with its multiplicity.

1.6   PROTECTION

Use duration of apparatus $oxy - CCR/Nx - SCR\ AMPHORA\ SCUBA$ depends on the *soda lime* canister CO_2 sorption duration and volume of breathing medium carried.

The assessment of soda lime sorption showed that the canister duration of $oxy - CCR\ AMPHORA\ SCUBA$ apparatus is within the range $\tau \in [150, 200]$ *min* (AQUA LUNG, 2004).

Use duration of $Nx - SCR\ AMPHORA\ SCUBA$ apparatus related to Nx volume carried was given earlier in Table 1.5. Duration capability of $oxy - CCR\ AMPHORA\ SCUBA$ apparatus related to the carried oxygen volume depends on the type of mission. Normally in the transit at a depth of above 6 mH_2O, the practical time of duration is up to 4 *h* (AQUA LUNG, 2004). Due to the way the oxygen is supplied, significant losses occur during the dives of a *yo-yo* type. During the descent there is an automatic refilling of oxygen to the breathing loop through dosage system 10 (Figure 1.3). During the ascent, the expanding breathing medium forces the diver to release the breathing medium through the mouth into water. Repetition of such an operation can cause a significant loss of the breathing medium thereby reducing the duration period of the *oxy – CCR AMPHORA SCUBA*.

Following the analysis presented earlier and taking into consideration the worst scenario, a conclusion can be made that the use duration of the diving apparatus is limited primarily by the regenerating capability of the CO_2 scrubber.

1.7 SUMMARY

This chapter describes some properties of the object of this study that have an impact on the safety of hyperbaric exposures, especially during dives in the semiclosed mode with *Nx* circulation used as breathing medium in *Nx – SCR AMPHORA SCUBA*.

Appropriate mathematical models of the apparatus's performance must be developed prior to experiments with humans. It is a mathematical model for ventilation of diving apparatus working in semiclosed circuit mode that is used for planning a safe decompression process. Such a model also provides the basis for the optimal design of diving equipment as well as determines the range of parameters necessary for its safe use.

An ability to plan effective purging of the breathing loop of a diving apparatus is indispensable to ensure its safe operation and effectiveness of the emergency procedures utilized to stabilize its work when the diver has doubts as to the composition of the inhaled breathing medium. It can be also used during emergency decompression. Since the Polish Naval Academy has already performed numerous experiments with similar types of apparatus, some tests can be omitted in the present study (Kłos R., 2000; Kłos R., 2021).

CO_2 scrubber use duration has the primary limiting effect on use duration of a diving apparatus. Thus, maximum time τ of a single mission using the *oxy – CCR/ Nx – SCR AMPHORA SCUBA* apparatus with oxygen during transit and for excursions to a maximum depth of 15 mH_2O, or *Nx* during trips from the transit depth to a maximum depth of 24 mH_2O, should be set at $\tau \leq 150$ *min*. In an emergency,[22] the time cannot exceed the limit $\tau < 200$ *min*.

For the proposed expertise, diving with the use of oxygen as a primary breathing medium with the possibility of *Nx* excursions, only one kind of premix has been recommended: *Nx* 0.43 with dosage rate $\dot{V} = 10.5\,dm^3 \cdot min^{-1}$ (Table 1.4). It made it possible to execute oxygen experimental dives with one *Nx* trip to the range of depths $H \in [6, 32]\ mH_2O$.

For the *Nx – SCR AMPHORA SCUBA* mode, the maximum theoretical duration of *Nx* trip is $\tau \leq 36$ *min*. While maintaining an emergency supply of *Nx* pressure at a level of $p = 5\ MPa$, this time is reduced to $\tau \leq 27$ min (Table 1.5).

NOTES

1 As marked in Figure 1.3.
2 Releasing the surplus of breathing medium into the water.
3 If it is open.
4 In accordance with a respiratory rhythm.

5 Depending on the planned maximum depth of dive.

6 Premix is a mixture of two or more constituents standardized before use; here this definition was used to underline the fact that Nx used for the supply of diving apparatus should be prepared beforehand, seasoned and undergone certification tests, and before use its composition should be operationally tested.

7 This will be shown later in this chapter in Tables 1.4–1.5.

8 Interchangeable orifices are fitted to the apparatus before a dive, and with the selected orifice a tank with properly chosen composition of premix is installed – this will be shown later in the chapter in Table 1.5.

9 After purified from carbon dioxide.

10 In fact, a small dependency from depth has been reported (Brennan D.M.A., Bdonchuk W.W., 1975; Kłos R., 2021).

11 *STP* conditions.

12 Absolute pressure at the maximum permitted operational depth foreseen for the diving system.

13 Absolute pressure at the minimal operational depth intended for the diving system.

14 Examples of these pressure values are quoted in Table 2.3.

15 Optimal values of premix contents and its dosage.

16 Of the same oxygen content.

17 Calculations were made for $Nx – SCR\ AMPHORA\ SCUBA$, but they bear the hallmarks of being general.

18 For example, through the nose.

19 Filling.

20 j represents the number of single purges.

21 Dead space has been evaluated here as 25% of the total volume $V_z = 0,25 \cdot V_c$

22 Forced by combat situation or to save human life, etc.

REFERENCES

AQUA LUNG. 2004. *AMPHORA Oxygen and Mix Gas Diving Equipment – Operating and Maintenance Manual*. Nice: La Spirotechnique I.C. AMPHORA Réf. A658.

Brennan DMA & Bdonchuk WW. 1975. Oxygen consumption of SCUBA divers – a technique for measurement and analysis. In *Proceedings of the 6-th International Conference on Underwater Education*. San Diego: National Association of Underwater Instructors.

Haux G. 1982. *Subsea Manned Engineering*. London: Bailliére Tindall. ISBN 0-7020-0749-8.

Kłos R. 2000. *Aparaty Nurkowe z regeneracją czynnika oddechowego*. Poznań: COOPgraf. ISBN 83-909187-2-2.

Kłos R. 2012. *Możliwości doboru ekspozycji tlenowo-nitroksowych dla aparatu nurkowego typu AMPHORA*. Gdynia: Polskie Towarzystwo Medycyny i Techniki Hiperbarycznej. ISBN 978-83-924989-8-8.

Kłos R. 2021. *Ventilation of Normobaric and Hyperbaric Objects*. Boca Raton, FL: CRC Press (Taylor & Francis Group, LLC). ISBN: 978-0-367-67523-3 (hbk) ISBN: 978-0-367-67524-0 (pbk) ISBN: 978-1-003-13164-9 (ebk).

Przylipiak M & Torbus J. 1981. *Sprzęt i prace nurkowe-poradnik*. Warszawa: Wydawnictwo Ministerstwa Obrony Narodowej. ISBN 83-11-06590-X.

Williams S. 1975. Engineering principles of underwater breathing apparatus. In *The Physiology and Medicine of Diving*, ed. PB Bennett & DH Elliott, 34–46. London: Baillière Tindall.

2 Inherent Unsaturation

Inert gas tensions in the tissues of a human body are in dynamic equilibrium with the components of air at atmospheric pressure. However, there is a difference in the dynamic equilibrium between arterial oxygen tension, venous tension and tissues tension. This phenomenon is often referred to as *inherent unsaturation* or *oxygen window*.

2.1 OXYGEN WINDOW

A significant part of oxygen from lungs enters the circulatory system, but it is only partially used up in metabolic reactions. The portion that takes part in the metabolic reactions[1] and reduces the dissolved oxygen tension[2] in the blood is referred to as *oxygen window*.[3] Comparing the levels of gas tension in venous circulation and in arteries, total tension difference under *STP* conditions is at a level of [8, 13]% (Kenny J.E., 1973). First, the oxygen reaction leading to the formation of water is responsible for the reduction of tensions of the gases in the body. Water formed under these conditions remains not only as water vapor but also as condensed matter.[4] Oxygen passes from the arterial blood into the tissues, where it is consumed, leaving the tension gap already mentioned. When breathing air at atmospheric pressure, the partial tension of oxygen in arterial blood is approximately $\pi_{O_2} \cong 100 \, mmHg$. During blood circulation, the level of oxygen tension decreases, and in venous capillaries it reaches approximately $\pi_{O_2} \cong 40 \, mmHg$[5] (Table 2.1).

Using oxygen as the breathing medium in the final phase of decompression can cause larger tension gap $\Delta \pi$ and in effect shorten the decompression time. This happens because oxygen helps flush out inert gases from the tissues and is metabolized relatively quickly there.

TABLE 2.1
Partial pressure/tension of the breathing gases (Przylipiak M., Torbus J., 1981).

Gas	Partial pressure/tension of the gas [*mmHg*]					
	Inhaled air	Air in alveoli	Arterial blood	Tissues	Venous blood	Exhaled air
Oxygen	158.0	100.0	95.0	40.0	40.0	116.0
Carbon dioxide	0.3	40.0	40.0	46.0	46.0	32.0
Nitrogen	596.0	573.0	573.0	573.0	573.0	565.0
Water vapor	5.7	47.0	47.0	47.0	47.0	47.0

DOI: 10.1201/9781003309505-3

2.2 MYOGLOBIN AND HEMOGLOBIN

An important moment in biochemistry was the discovery of the structures of two proteins: *myoglobin* and *hemoglobin*. The main function of myoglobin is to store oxygen in striated muscles,[6] while that of hemoglobin is to store and transfer oxygen through blood. Storage of oxygen by myoglobin *Mb* is well described by mathematical model proposed by *Leonor Michaelis* and *Maud Menten*. Their model describes kinetics of an enzymatic reaction. Its algebraic form is known as the *Michaelis-Menten equation*. The process of oxygen storage involves an intermediate step of forming a complex of oxygen with myoglobin MbO_2 followed by release of oxygen to tissues O_2^* and restoration of myoglobin molecule *Mb*:

$$Mb + O_2 \underset{k_1'}{\overset{k_1}{\rightleftharpoons}} MbO_2 \overset{k_2}{\rightarrow} Mb + O_2^* \qquad (2.1)$$

where: *Mb* – myoglobin, O_2^* – oxygen released to tissue.

The existence of transition state[7] is confirmed by the saturation of *Mb* in the presence of a high concentration of O_2,[8] as resulting from the use of all active sites in the structure of Mb[9] (Figure 2.1).

Other MbO_2 complexes can be created only after the breakdown of the existing ones. The existence of the MbO_2 complex was shown also by other methods (Stryer L., 1997; Berg J.M., Tymoczko J.L., Stryer L., 2013).

Later in the text we use relevant concentrations of the compounds in question in square brackets: [*Mb*] – concentration of myoglobin, [O_2] – oxygen concentration, [MbO_2] – myoglobin-oxygen complex concentration.

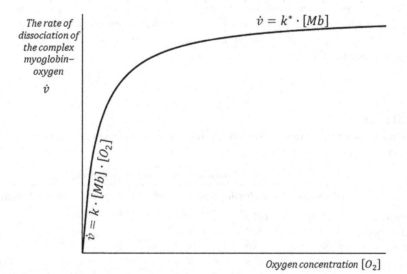

FIGURE 2.1   The dependence of the rate of dissociation $\dot{v}$ of the complex myoglobin-oxygen versus oxygen concentration [O_2].

In the Michaelis-Menten model shown in Figure 2.1, it can be seen that in the initial stage, for low concentrations of oxygen $[O_2]$ reaction rate, $\dot{v}$ is proportional to the concentration of myoglobin $[Mb]$ and oxygen concentration $[O_2]$: $\dot{v} = k \cdot [Mb] \cdot [O_2]$, where: k is the rate constant $[dm^6 \cdot mol^{-1} \cdot s^{-1}]$. At higher concentrations of oxygen $[O_2]$, because of its excess, the rate of reaction $\dot{v}$ depends only on the concentration of myoglobin $[Mb]$: $\dot{v} = k^* \cdot [Mb]$, where k^* is the maximum rate constant for reaction $[dm^3 \cdot s^{-1}]$.

According to the assumptions of model (2.1), the MbO_2 complex formation reaction is followed by its dissociation, transfer of oxygen O_2^* to the tissues and the restoration of the Mb myoglobin molecules. The rate of individual reactions $\dot{v}$ for total chemical reaction (2.1) can be written as $\dot{v}_1 = k_1 \cdot [Mb] \cdot [O_2]$, $\dot{v}_1' = k_1' \cdot [MbO_2]$ and $\dot{v}_2 = k_2 \cdot [MbO_2]$, where k are the reaction rate constants in accordance with the previously accepted designation for total chemical reaction (2.1).

The total rate of myoglobin binding can be written as $-\frac{\partial[Mb]}{\partial t} = \dot{v}_1 - \dot{v}_1' = k_1 \cdot [Mb] \cdot [O_2] - k_1' \cdot [MbO_2]$, where t is time. The transfer rate of oxygen to the tissue cells as indicated in (2.1) is $\dot{v}_2 = k_2 \cdot [MbO_2]$ and hence the rate of change of the concentration of the $[MbO_2]$ complex can be written as $-\frac{\partial[MbO_2]}{\partial t} = \dot{v}_1 - \dot{v}_1' - \dot{v}_2 = k_1 \cdot [Mb] \cdot [O_2] - k_1' \cdot [MbO_2] - k_2 \cdot [MbO_2]$. For the steady state, the rate of change of the concentration of the $[MbO_2]$ complex will be zero $-\frac{\partial[MbO_2]}{\partial t} = 0$ and the total content of myoglobin $[Mb]_0$ in the steady state is the sum of concentrations of $[MbO_2]$ complexes and free $[Mb]$: $[Mb]_0 = [MbO_2] + [Mb]$. Using these calculations, the $[MbO_2]$ complex concentration may be written as:

$$[MbO_2] = \frac{k_1 \cdot [O_2]}{k_2 + k_1' + k_1 \cdot [O_2]} \cdot [Mb]_0 \qquad (2.2)$$

where: $[MbO_2]$ – molar concentration of oxidized form of myoglobin, $[Mb]_0$ – total molar concentration of myoglobin in both the bound form and in the free form, $[O_2]$ – molar oxygen concentration, k_1 – the rate of formation of the bounded form of myoglobin MbO_2, k_1' – the rate of dissociation reaction of myoglobin bound form MbO_2, k_2 – reaction rate of supply of the oxygen to the tissue cells from the bound form of myoglobin MbO_2.

Substituting equation (2.2) for the relationship describing the rate of oxygen transfer to tissue: $\dot{v}_2 = k_2 \cdot [MbO_2]$, one can obtain an algebraic model of the reaction rate of oxygen donation by myoglobin: $\dot{v}_2 = \dot{v} = \frac{k_1 \cdot k_2 \cdot [O_2]}{k_2 + k_1' + k_1 \cdot [O_2]} \cdot [Mb]_0$. After dividing the numerator and denominator by $k_1 \cdot [O_2]$, and substituting $K_M = \frac{k_2 + k_1'}{k_1}$, one can write: $\dot{v}_2 = \dot{v} = \{k_2 \cdot [Mb]_0\} : \left\{ \frac{K_M}{[O_2]} + 1 \right\}$. K_M value is referred to as the Michaelis constant.[10] When the K_M value is much lower than the value of oxygen concentration $[O_2] \gg K_M$, the fraction of $\frac{K_M}{[O_2]}$ tends to zero and rate $\dot{v}_2$ will reach, in this case, the maximum value: $\dot{v}_{max} = k_2 \cdot [Mb]_0$, because all the myoglobin is bound to oxygen in MbO_2 complex. Using the previous, one can write the Michaelis-Menten equation when applied to modelling oxygen transfer process from Mb as:

$$\exists_{[O_2] \gg K_M} \quad \dot{v} = \frac{\dot{v}_{max}}{\frac{K_M}{[O_2]}+1} \Rightarrow \frac{\dot{v}}{\dot{v}_{max}} = \frac{[O_2]}{K_M + [O_2]} \tag{2.3}$$

where: $\dot{v}$ – rate of oxygen transfer process, $\dot{v}_{max}$ – maximum rate of oxygen transfer process, $[O_2]$ – oxygen concentration, K_M – Michaelis constant.

2.3 THE AFFINITY FOR OXYGEN AND THE DEGREE OF SATURATION

The degree of saturation of myoglobin with oxygen x_{O_2} can be defined as the ratio of concentration of $[MbO_2]$ to the total concentration of myoglobin $[Mb]_0 = [MbO_2] + [Mb]$: $x_{O_2} = \frac{[MbO_2]}{[MbO_2]+[Mb]}$. In line with the analysis of the problem, the degree of saturation x_{O_2} will be proportional to the ratio of the rate of reaction of oxygen binding $\dot{v}$ by Mb to its maximum value $\dot{v}_{max}$: $x_{O_2} = \frac{\dot{v}}{\dot{v}_{max}}$. In order to simplify further analysis, the so-called Mb affinity for oxygen[11] P_{50} should be defined as oxygen partial pressure π_{O_2} at which the saturation of myoglobin x_{O_2} is expected to be equal to half the maximum value $x_{O_2} = 0.5\,mol \cdot mol^{-1}$: $P_{50} \equiv \pi_{O_2}(x_{O_2} = 0.5)$. By multiplying the numerator and denominator of the right-hand side of equation (2.3) by tension π of the total blood gases, one can write: $\frac{\dot{v}}{\dot{v}_{max}} = \frac{\pi \cdot [O_2]}{\pi \cdot K_M + \pi \cdot [O_2]} = \frac{\pi_{O_2}}{\pi \cdot K_M + \pi_{O_2}}$.

Assuming $\dot{v}_{max} = 2 \cdot \dot{v}$, degree of saturation x_{O_2} for Mb is $x_{O_2} = \frac{\dot{v}}{\dot{v}_{max}} = \frac{\dot{v}}{2 \cdot \dot{v}} = 0.5$.

Using the definition of affinity P_{50} one can write: $\frac{P_{50}}{\pi \cdot K_M + P_{50}} = 0.5 \Rightarrow \pi \cdot K_M = P_{50}$.

Thus, the Michaelis constant K_M according to (2.3) representing myoglobin's affinity for oxygen[12] can be replaced by value P_{50} (Mb) ← K_M of the oxygen affinity for myoglobin when oxygen concentration $[O_2]$ will be replaced by its partial tension $\pi_{O_2} \leftarrow [O_2]$:

$$\exists_{\pi_{O_2} \gg P_{50}} \quad x_{O_2} = \frac{\pi_{O_2}}{\pi_{O_2} + P_{50}} \tag{2.4}$$

where: P_{50} – affinity of Mb for oxygen, x_{O_2} – myoglobin saturation with oxygen, π_{O_2} – oxygen tension.

The algebraic model of oxygen saturation of myoglobin[13] Mb (2.4) is represented by a hyperbole.[14] This model was validated experimentally by determining the oxygen affinity P_{50} for myoglobin $P_{50}(Mb) = 1$ $mmHg$, with close overlapping of the theoretical model with the experimental curve (Stryer L., 1997; Berg J.M., Tymoczko J.L., Stryer L., 2013).

2.4 HILL'S SIGMOIDAL MODEL

Archibald Hill postulated that the process of oxygen binding to hemoglobin Hb can be written in the form of reactions:

$$Hb + n \cdot O_2 \underset{k_1'}{\overset{k_1}{\rightleftharpoons}} Hb(O_2)_n \overset{k_2}{\rightarrow} Hb + n \cdot O_2^* \tag{2.5}$$

which leads to other than (2.3) form of mathematical model, called the *Hill equation*:

$$\exists_{[O_2] \gg K} \frac{\dot{v}}{\dot{v}_{max}} = \frac{[O_2]^n}{K + [O_2]^n} \quad \text{(Table 2.2)}.$$

TABLE 2.2
Derivation of the Hill equation according to the reaction (2.5)

Specification	Equation
Rate of oxygen binding by Hb	$\dot{v}_1 = k_1 \cdot [Hb] \cdot [O_2]^n$
Dissociation of $Hb(O_2)_n$ complex	$\dot{v}_1' = k_1' \cdot [Hb(O_2)_n]$
Transfer of oxygen to cells through complex $Hb(O_2)_n$	$\dot{v}_2 = k_2 \cdot [Hb(O_2)_n]$
Total rate of binding of Hb	$-\dfrac{\partial [Hb]}{\partial t} = \dot{v}_1 - \dot{v}_1' = k_1 \cdot [Hb] \cdot [O_2]^n - k_1' \cdot [Hb(O_2)_n]$
Total rate of concentration change of complex $Hb(O_2)_n$	$-\dfrac{\partial [Hb(O_2)_n]}{\partial t} = \dot{v}_1 - \dot{v}_1' - \dot{v}_2 = k_1 \cdot [Hb] \cdot [O_2]^n - k_1' \cdot [Hb(O_2)_n]$ $\qquad\qquad - k_2 \cdot [Hb(O_2)_n]$
For steady state	$-\dfrac{\partial [Hb(O_2)_n]}{\partial t} \equiv 0 \wedge [Hb]_0 \cdot [Hb(O_2)_n] + [Hb]$ $[Hb(O_2)_n] = \dfrac{k_1 \cdot [O_2]^n}{k_1' + k_2 + k_1 \cdot [O_2]^n} \cdot [Hb]_0$
Rate of oxygen transfer to cells by Hb(O$_2$)$_n$ complex	$\dot{v}_2 = k_2 \cdot [Hb(O_2)_n] = k_2 \cdot \dfrac{k_1 \cdot [O_2]^n}{k_1' + k_2 + k_1 \cdot [O_2]^n} \cdot [Hb]_0 / k_1 \cdot [O_2]^n$

(Continued)

TABLE 2.2 (*Continued*)
Derivation of the Hill equation according to the reaction (2.5)

Specification	Equation

$$\dot{v}_2 = \frac{k_2}{\dfrac{k_1' + k_2}{k_1 \cdot [O_2]^n} + k_1 \cdot [O_2]^n} \cdot [Hb]_0 \quad \text{introduce}: K \equiv \frac{k_1' + k_2}{k_1}$$

$$\dot{v}_2 = \frac{k_2}{\dfrac{K}{[O_2]^n} + 1} \cdot [Hb]_0$$

Condition for maximization of the oxygen transfer rate of

$$\lim_{K/[O_2]^n \to 0} \dot{v}_2 = \dot{v}_{max} = k_2 \cdot [Hb]_0$$

Rate of oxygen transfer to cells by $Hb(O_2)_n$ complex

$$\dot{v} \equiv \dot{v}_2 = \frac{\dot{v}_{max} \cdot [O_2]^n}{K + [O_2]^n}$$

Hill equation can be similarly written as a function of oxygen saturation $x_{O_2} \leftarrow \dfrac{\dot{v}}{\dot{v}_{max}}$, of hemoglobin, partial tension of oxygen $\pi_{O_2} \leftarrow [O_2]$ and the hemoglobin affinity for oxygen $P_{50}^n (Hb) \leftarrow K$ in the form[15]: $\exists_{\pi_{O_2} \gg P_{50}} x_{O_2} = \dfrac{\pi_{O_2}^n}{\pi_{O_2}^n + P_{50}^n}$ or equivalently in the form of:

$$\exists_{\pi_{O_2} \gg P_{50}} \frac{x_{O_2}}{1 - x_{O_2}} = \left(\frac{\pi_{O_2}}{P_{50}}\right)^n \Leftrightarrow x_{O_2} = \frac{\pi_{O_2}^n}{\pi_{O_2}^n + P_{50}^n} \tag{2.6}$$

The algebraic model of oxygen saturation of Hb[16] (2.6) has a sigmoidal shape and was validated experimentally by determining the hemoglobin affinity for oxygen $P_{50}(Hb) = 26 \, mmHg$ and a *Hill coefficient*[17] $n_{Hb} = 2.8$ (Stryer L., 1997; Ekeloef N.P., Eriksen J., Kancir C.B., 2001; Berg J.M., Tymoczko J.L., Stryer L., 2013). The Hill coefficient for myoglobin Mb is $n_{Mb} = 1$ (Stryer L., 1997; Berg J.M., Tymoczko J.L., Stryer L., 2013).

The graphs showing algebraic models of dissociation of myoglobin–oxygen and hemoglobin–oxygen complexes are shown in Figure 2.2.

The higher value of the Hill coefficient for Hb binding is known as cooperativity of binding O_2 by Hb. It means that O_2 that is associated with one hem facilitates the binding of another O_2 in the same hem tetramer (Berg J.M., Tymoczko J.L., Stryer L., 2013). If we assume that Mb takes part in oxygen transport in the blood similarly as Hb, and compare Hb saturation with O_2 as a function of oxygen partial tension, it

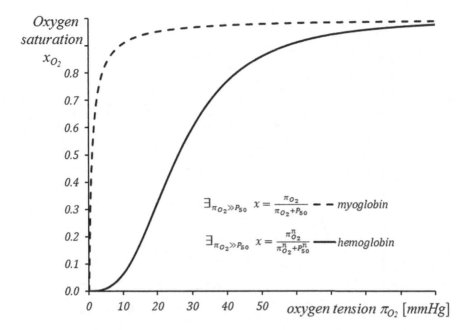

FIGURE 2.2 Dissociation models of oxygen–myoglobin and oxygen–hemoglobin complexes relating oxygen saturation $x_{O_2} = \frac{[AO_2]}{[AO_2]+[A]} \big| A = \{Hb; Mb\}$ to oxygen tension in the blood π_{O_2}.

will mean a greater change than for Mb. Although in reality, Mb will be saturated to a greater extent than Hb[18] (Figure 2.3).

2.5 PHYSICAL SOLUBILITY OF OXYGEN IN BLOOD

Under normal conditions, nearly the entire amount of oxygen transported by blood is present in a complex with hemoglobin Hb[19] (Stryer L., 1997; Berg J.M., Tymoczko J.L., Stryer L., 2013). Only a small part of it is dissolved in the blood,[20] but it plays a key role in the diffusion-related oxygen transport mechanism to cells. Physical solubility $R(p)$ of oxygen in the blood as a function of pressure can be written approximately as $R(p) \cong 3 \cdot 10^{-2} \, cm^3 O_2 \cdot mmHg^{-1} \, O_2 \cdot dm^{-3}$ of blood. From equation (2.6) for oxygen saturation $x_{O_2} = 0.97 \Rightarrow \pi_{O_2} = 89.5 \, mmHg$. The concentration of physically dissolved oxygen will be approximately $R \cong R(p) \cdot \pi_{O_2} \cong 2.7 \, cm^3 \, O_2 \cdot dm^{-3}$ of blood (Figure 2.4).

2.6 TRANSPORT OF OXYGEN

Like myoglobin Mb for the muscle tissue, hemoglobin Hb for the blood is an oxygen repository. Reducing the oxygen concentration physically dissolved in the blood causes release of oxygen from hemoglobin Hb.

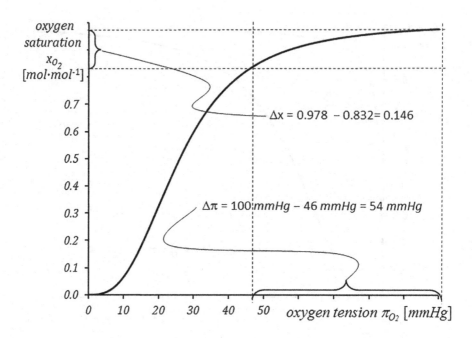

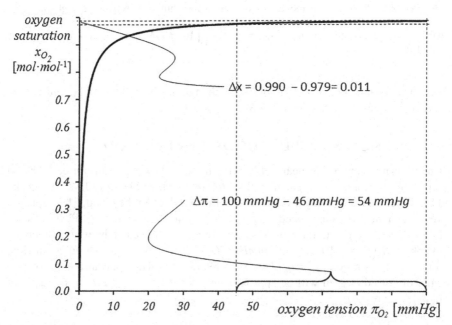

FIGURE 2.3 Dissociation models of (a) oxygen–hemoglobin and (b) oxygen–myoglobin complexes relating oxygen saturation $x_{O_2} = \dfrac{[AO_2]}{[AO_2]+[A]} \mid A = \{Hb; Mb\}$ to oxygen tension in the blood π_{O_2}, assuming that the oxygen partial pressure in the alveoli will be at $p_{O_2} = 100\,mmHg$ and oxygen tension in the capillaries $\pi_{O_2} = 46\,mmHg$.

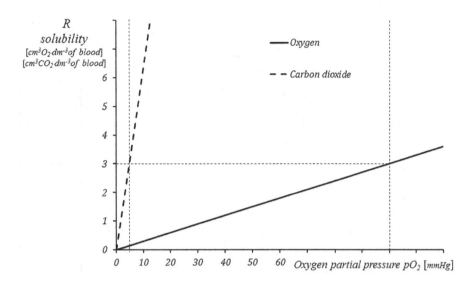

FIGURE 2.4 Comparison of physical solubility R of oxygen and carbon dioxide in the blood as a function of oxygen partial pressure p_{O_2} in the inhaled respiratory medium.

Under normal conditions, the maximum saturation of hemoglobin Hb is about $Y_{Hb} \cong 1.39 \, cm^3 \, O_2 \cdot g^{-1} \, Hb$. In a healthy person the average hemoglobin content C_{Hb} in $1 \, dm^3$ of blood is approximately $C_{Hb} \cong 150 \, g \, Hb \cdot dm^{-3}$ of blood.

Values C_{Hb} and Y_{Hb} were used to calculate the additional graph scale (Figure 2.5) for concentration C of oxygen bound to Hb in the blood.[21]

At rest, in a sitting position blood flow $\dot{V}$ is at a level $\dot{V} \in [4,6] \, dm^3 \cdot min^{-1}$ with the consumption of oxygen by the tissue at the level of $\dot{v} \in [13,100] \, cm^3 \cdot min^{-1}$ (Table 2.4).

The averaged oxygen consumption rate $\dot{v}_{O_2}$ for the human body is at the level $\dot{v}_{O_2} \in [0.2, 0.3] \, dm^3 \cdot min^{-1}$ (Åstrand P.-O., Rodahl K., 1977).

With an increase in the workload, the consumption rate rises (Table 2.3).

Under normal pressure in pulmonary arterioles, oxygen tension is approximately $\pi_{O_2} \cong 100 \, mmHg$. Under workload, the oxygen tension in muscle capillaries is around $\pi_{O_2} \cong 20 \, mmHg$ (Lamberdsen C.J., Kough R.H., Cooper D.Y., Emmel G.L., Loeschcke, Schmidt C.F., 1953). This example corresponds to a decrease in oxygen tension for a consumption rate at a level of approximately $\dot{v}_{O_2} \cong 0.7 \, dm^3 \cdot min^{-1}$ for light exercise (Figure 2.5).

2.7 OXYGEN CONSUMPTION

In Table 2.3, the most common global oxygen consumption rates $\dot{v}_{O_2}$ are listed, but at the level of individual organs of a body, the oxygen consumption rate

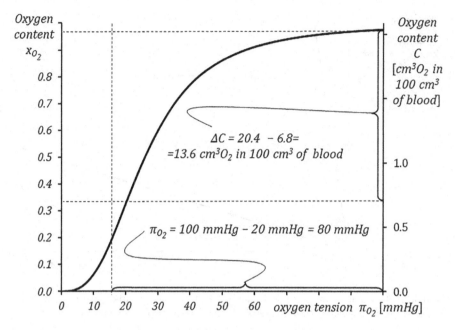

FIGURE 2.5 Oxygen content difference between pulmonary arteries and capillaries corresponding to the decrease in oxygen tension $\Delta\pi_{O_2}$ in the blood for light work[22] as in Table 2.3.

TABLE 2.3
Rate of oxygen consumption and lung ventilation as a function of physical effort (Przylipiak M., Torbus J., 1981; Kłos R., 2021).

Physical effort		Stream of consumer oxygen	Breaths per minute	Lung ventilation	Border stream of consumer oxygen
Intensity	Example	$[dm^3 \cdot min^{-1}]$	$[min^{-1}]$	$[dm^3 \cdot min^{-1}]$	$[dm^3 \cdot min^{-1}]$
very light	lying in bed	0.25	to 20	8–10	to 0.5
	sitting	0.30			
	motionless standing	0.40			
light	walk 3.5 $km \cdot h^{-1}$	0.7	20–25	10–20	0.5–1.0
moderate	march 6.5 $km \cdot h^{-1}$	1.2	25–30	20–30	1.0–1.5
heavy	swimming with speed of 3.0 $km \cdot h^{-1}$	1.8	30–35	30–50	1.5–2.0
very heavy	running with speed of 13 $km \cdot h^{-1}$	2.0	35–40	50–65	2.0–2.5
extremely heavy	uphill running	4.0	>40	>65	>2.5

TABLE 2.4
Amount of oxygen drawn from the blood into the tissues (Vann R.D., 1989).

Body part	Intake of oxygen from blood for tissue level
	[$cm^3 \cdot dm^{-3}$ of blood]
Heart	100
Brain	60
Digestive system	60
Muscles at rest	50
Kidneys	13
Other	50

varies quite widely (Table 2.4). Analyzing Figures 2.2–2.3 and Figure 2.5, one can conclude that the higher the rate of consumption of oxygen, the wider the oxygen window.

It does not translate directly into the risk of oxygen toxicity *CNSyn* for various organs, since their resistance to harmful metabolites produced also varies significantly. Thus, despite the fact that the nervous system uses a fairly large amount of oxygen as compared to other organs, its oxygen sensitivity is also higher. This means that the size of oxygen window cannot be used for *CNSyn* hazard assessment.

2.8   THE ENVIRONMENTAL IMPACTS

Qualitative dependence of the degree of saturation x_{O_2} on hemoglobin HbA[23] as function of oxygen tension π_{O_2} and pH[24] is shown in Figure 2.6 (Corda M., De Rosa M.C., Pellegrini M.G., Sanna M.T., Olianas A., Fais A., Manca L., Masala B., Zappacosta B., Ficarra S., Castagnola M., Giardina B., 2000).

This dependence shows that an increase in activity of hydronium ions[25] $a_{H_3O^+}$ reduces *HbA* affinity for oxygen, causing easier oxygen donation.[26] Increased *pH* produces a rise in affinity of *HbA* and makes it difficult to transfer oxygen to tissues. This evolution, for each type of Hb, is the same as for *HbA*. This phenomenon is called the *Bohr effect*. In the body, this effect is caused most commonly by carbonic acid formed from CO_2 dissolved in blood, which under the influence of carbonic anhydrase decomposes into hydronium cation H_3O^+ and bicarbonate anion HCO_3^-:

$$H_2CO_3 \quad \overset{carbonic\,anhydrase}{\longrightarrow} \quad H_3O^+ + HCO_3^-$$

Aside from the impact of CO_2 content, the temperature also causes a change in affinity of hemoglobin to oxygen (Figure 2.7).

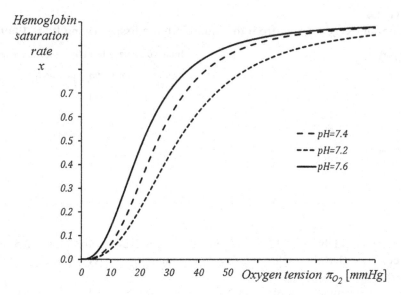

FIGURE 2.6 Relation of hemoglobin saturation rate x with low affinity to oxygen *HbA* as oxygen tension function π_{O_2} and *pH* blood (Corda M., De Rosa M.C., Pellegrini M.G., Sanna M.T., Olianas A., Fais A., Manca L., Masala B., Zappacosta B., Ficarra S., Castagnola M., Giardina B., 2000).

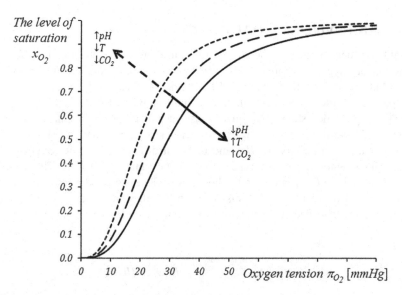

FIGURE 2.7 Changes in the level of saturation x_{O_2} of hemoglobin *Hb* as a function of oxygen tension π_{O_2} in the blood, *pH*, temperature T and CO_2 content.

2.9   HALDANE EFFECT

The *Haldane effect* involves easier CO_2 binding by reduced hemoglobin than in the exchange reaction of oxy-hemoglobin and is linked to the Bohr effect, which is a phenomenon of a decrease in hemoglobin *Hb* affinity for oxygen with a decrease in *pH* of blood, as discussed earlier (Figure 2.7).

Carbon dioxide is transferred by blood from tissues to lungs in several different ways. Approximately 60% of CO_2 is transported in the form of bicarbonate ions[27] formed from water and CO_2 in the presence of an enzyme – carbonic anhydrase.

About [5, 30]% CO_2 is transported as *carbomino–hemoglobin* complex CO_2–Hb.[28]

Transport by means of binding to hemoglobin disappears when breathing pure oxygen at a pressure of about 300 *kPa*, because in this case *Hb* is almost completely blocked by oxygen and metabolic reaction uses only the oxygen physically dissolved in the blood (Nunn J.F., 1993; West J.B., Luks A.M., 2016). On this basis it was concluded that the disruption of CO_2 transfer taking place in this way is accompanied by its retention, which causes symptoms of oxygen poisoning. However, it seems that the mechanism is different[29] (Lamberdsen C.J., Kough R.H., Cooper D.Y., Emmel G.L., Loeschcke H.H., Schmidt C.F., 1953).

About 10% CO_2 is transported by plasma in the form of plasma protein carbamic compounds. Carbamates are the result of direct bonding of CO_2 to amino groups without CO_2 hydration, and the actual CO_2 vapor pressure has little effect on the amount of CO_2 transported in the form of carbamates.

2.10   RISK OF DEVELOPING CENTRAL NERVOUS SYNDROME

Central nervous syndrome (*CNSyn*) is a neurological condition caused by damage to or dysfunction of the *central nervous system* (*CNS*), which includes the brain, brainstem and spinal cord.[30]

From experimental measurements it is known that an increase in the level of oxygen partial pressure in a breathing medium to approximately $p_{O_2} \cong 345\,mmHg$ may cause the arterial blood oxygen tension to rise approximately to $\pi_{O_2} \cong 300\,mmHg$, and in venous capillaries it will remain at the level of approximately $\pi'_{O_2} \cong 40\,mmHg$ or will slightly exceed it. Considering $\pi_{CO_2} \cong 7\,mmHg$ as CO_2 tension, it can be assumed that the total tension of O_2 and CO_2 in veins will be about $\pi_{CO_2+O_2} \cong 50\,mmHg$. Subtracting this value from the oxygen tension in arterial blood π_{O_2}, one can estimate the value of the oxygen window at $\Delta\pi \cong 300 - 50 = 250\,mmHg$. This tension deficiency can be made available for other gases, which should leave the body in the process of decompression (Behnke A.R., 1967).

In the *SEA – LAB II* experiment at a depth of 200 *fsw*,[31] in which divers breathed a medium containing 4.5% O_2[32] for 14 days, the average oxygen tension in arterial blood was $\pi_{O_2} \cong 192\,mmHg$, while in the veins it was $\pi'_{O_2} \cong 40\,mmHg$ (Behnke A.R., 1967). From these data and from equation (2.6), saturation x_{O_2} can be estimated for arterial oxygen tension $\pi_{O_2} = 192\,mmHg\,O_2$ at a level $x_{O_2}\left(192\,mmHg\,O_2\right) \cong 0.996$. And for the venous blood, for which the oxygen tension was at a level of $\pi_{O_2} = 40\,mmHg\,O_2$, it would be only $x'_{O_2}\left(40\,mmHg\,O_2\right) \cong 0.770$. This corresponds to the volume of oxygen

x_v bound by arterial blood hemoglobin Hb at a level $x_v = 0.2085\ dm^3\ O_2 \cdot dm^{-3}$ of blood $0.996 \cong 0.208\ dm^3\ O_2 \cdot dm^{-3}$ of blood.[33]

Similarly, for the transport of oxygen by venous blood hemoglobin Hb, the content of transported oxygen can be determined at a level of approximately $x_v' \cong 0.161\ dm^3 O_2 \cdot dm^{-3}$ of blood. The content of oxygen physically dissolved in arterial blood is $x_v = 3 \cdot 10^{-5}\ dm^3\ O_2 \cdot mmHg^{-1}\ O_2 \cdot dm^{-3} \cdot 192\ mmHg \cong 0.006 dm^3 O_2 \cdot dm^{-3}$ of blood, and in venous blood $x_v' \cong 3 \cdot 10^{-5}\ dm^3 O_2 \cdot mmHg^{-1}\ O_2 \cdot dm^{-3}$ of blood $\cdot 40\ mmHg \cong 0.001\ dm^3 O_2 \cdot dm^{-3}$ of blood.

Altogether there is $x_{xy} \cong 0.214\ dm^3 O_2 \cdot dm^{-3}$ of blood transferred through arteries and through veins $x_v \cong 0.162\ dm^3 O_2 \cdot dm^{-3}$ of blood. The difference between the oxygen content in one liter of arterial and venous blood is $\Delta V_{O_2} \cong 0.052\ dm^3 O_2$ under normal conditions.

For normal conditions and the oxygen consumption within $\dot{v}_{O_2} \in (0.5, 3.0)\ dm^3 O_2 \cdot min^{-1}$, blood flow $\dot{V}$ is an approximate linear function of the oxygen consumption $\dot{v}_{O_2}$: $\dot{V} = f\left(\dot{v}_{O_2}\right)$, which can be written as $\dot{V}\left(\dot{v}_{O_2}\right) \cong 6\ dm^3$ of blood $\cdot dm^{-3} O_2 \cdot \dot{v}_{O_2} + 2 dm^3$ of blood $\cdot min^{-1}$ (Åstrand, P.-O., Rodahl, K., 1977). Below the minimum oxygen consumption of approximately $\dot{v}_{O_2} < 0.5\ dm^3 O_2 \cdot min^{-1}$ stabilizes at around $\dot{V} \cong 5 dm^3$ of blood $\cdot min^{-1}$.

Bradycardia hinders the estimation of oxygen consumption $\dot{v}_{O_2}$ as a function of pulse.[34] The average consumption of oxygen $\dot{v}_{O_2}$ during the $SEA - LAB\ II$ experiment can be estimated at a level of $\dot{v}_{O_2} \cong 0.25\ dm^3 O_2 \cdot min^{-1}$ for an estimated blood flow $\dot{V}$ at a level of $\dot{V} \cong 5 dm^3$ of blood $\cdot min^{-1}$, which according to Table 2.3 represents light effort.

In Table 2.5 and Figure 2.8, one can see the direction of change in oxygen content C_{O_2} as a function of its tension π_{O_2} for very light effort[35] (Lamberdsen C.J., Kough R.H., Cooper D.Y., Emmel G.L., Loeschcke H.H., Schmidt C.F., 1953).

Under normal conditions, the maximum saturation of 1 g of Hb is approximately $1.39\ cm^3 O_2$. A healthy person's blood, on average, contains 1 dm^3 of blood in approximately 150 g of Hb, hence according to (2.6) O_2 content bound in a Hb will be: $x_v = 1.39\ cm^3 O_2 \cdot g^{-1} \cdot 150\ g \cdot dm^{-3}$ of blood x_{O_2}.

Physical solubility of oxygen in the blood of $R(p)$ is approximately $R(p) \cong 3 \cdot 10^{-2}\ cm^3 O_2 \cdot mmHg^{-1}\ O_2 \cdot dm^{-3}$ of blood, hence the physical content magnitudes of oxygen dissolved in blood can be written as the following formula: $x_v = 3 \cdot 10^{-2}\ cm^3 O_2 \cdot mmHg^{-1}\ O_2 \cdot dm^{-3} \cdot \pi_{O_2}$. The solid line $C_{O_2} = x_v \cdot 100\%$ in Figure 2.8 is the sum[36] of the oxygen content bound to hemoglobin C_{Hb} and dissolved physically C: $C_{O_2} = C_{Hb} + C$. The points represent the average values and are marked as follows: T represents arterial blood, Z refers to venous blood, index 1 refers to breathing air under normal conditions, index 2 refers to breathing oxygen under normal conditions, index 3 refers to breathing pure oxygen at a pressure of 2660 $mmHg$[37] and index 4 refers to results of the $SEA - LAB\ II$ experiment (Table 2.5).

When arterial blood passes through tissues nearly $6\%_v O_2$ enters into biochemical reactions in cells. The reactions occurring lead primarily to formation of H_2O and CO_2. The consumption of oxygen[38] is dependent only on the demand for the energy stored in the cell as bonds[39] within the guanosine triphosphate (GTP) and adenosine triphosphate (ATP) molecules. It does not depend on the partial pressure of oxygen p_{O_2} in the inhaled breathing medium.[40]

TABLE 2.5
Tension and content values of oxygen (Lamberdsen C.J., Kough R.H., Cooper D.Y., Emmel G.L., Loeschcke H.H., Schmidt C.F., 1953).

Breathing medium	Pressure		Oxygen initial tension Oxygen contents		
	Total	Partial of oxygen	Aorta	Vena cava	Difference
	[mmHg]	[mmHg]	$\begin{bmatrix} mmHg \end{bmatrix}$ $[\%_v]$	$\begin{bmatrix} mmHg \end{bmatrix}$ $[\%_v]$	$\begin{bmatrix} mmHg \end{bmatrix}$ $[\%_v]$
Oxygen	760	760	130^* 21.0	40 13.4	90 7.6
	2660	2660	2100 26.0	75 17.8	2025 8.2
Air	760	160	91 18.7	55.7 12.6	35,3 6.1
Heliox	5320	239	192 20.8^*	40 16.1^*	152 4.7^*

* calculated values

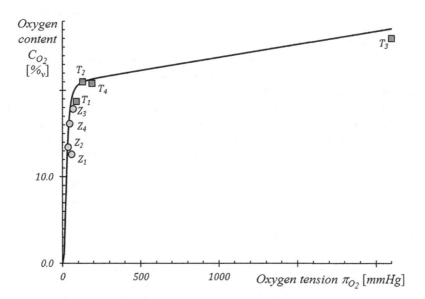

FIGURE 2.8 The content of oxygen C_{O_2} as a function of its tension π_{O_2} in the blood, where the solid line shows the values resulting from the model used and described in the text, and the points represent the practical values – Table 2.5 (Lamberdsen C.J., Kough R.H., Cooper D.Y., Emmel G.L., Loeschcke H.H., Schmidt C.F., 1953).

When under normal conditions air is breathed and light effort is made, the oxygen tension difference in humans $\Delta\pi_{O_2}$ between arterial blood and a venous blood is at a level $\Delta\pi_{O_2} \cong 49\,mmHg$.[41] Point $\hat{T}_1$ represents oxygen tension π_{O_2}, while breathing air in normal conditions. Under these conditions, Hb is almost completely saturated with oxygen; theoretically, the degree of saturation is $x_{O_2} \cong 0.971$.

The earlier mentioned consumption of approximately $6\%_v O_2$ in metabolic reactions theoretically leaves saturation of Hb at a level of $x_{O_2} \cong 0.894$. In the course of breathing O_2 under an increased pressure, its continuously increasing portion is transmitted through plasma, while Hb stays bonded[42] all the time to O_2. When the value of partial oxygen tension is around $\pi_{O_2} \cong 150\,mmHg$ the increase in C_{O_2} is noticed. It is linked only to its physical solubility in the blood following the increase in oxygen partial pressure p_{O_2} in the inhaled breathing medium. Above tension value $\pi_{O_2} \cong 150\,mmHg$, a decrease in the oxygen window width is noticed together with an increase in C_{O_2} physically dissolved in the blood above the typical value of $C_{O_2} > 3\cdot10^{-3}\,cm^3 O_2 \cdot 100\,cm^{-3}$ of blood (Figure 2.8). This way, the tissues are exposed to increasingly higher oxygen partial tensions π_{O_2}, interacting with arterial blood which is forced to flow through them. Those tensions can be regarded as directly proportional to the risk of creation of highly oxidized metabolites and radicals, which can cause $CNSyn$ symptoms, as described by the theory of biochemical oxygen poisoning in Chapter 3.

In simple terms, it can be assumed that the oxygen partial pressure p_{O_2} in the inhaled breathing medium exceeded to the level of complete saturation of hemoglobin Hb in arterial blood constitutes the limit to the protective effect produced by the oxygen window associated with oxygen partial tension π_{O_2} which physically saturates blood at a safe level. From this moment, the period of breathing with atmosphere enriched with O_2 is limited.

Breathing O_2 is accompanied by the effect of increased resistance of cerebral blood flow initially limiting the effect of exposure of the nerve cells to oxygen stream O_2 physically dissolved in blood (Lamberdsen C.J., Kough R.H., Cooper D.Y., Emmel G.L., Loeschcke H.H., Schmidt C.F., 1953). When breathing oxygen under pressure $p_{O_2} = 350\,kPa$, the resistance of cerebral vessels increases by half, causing about 25% reduction in blood flow through the brain. This is roughly a double effect when comparing reduction of cerebral blood flow accompanying breathing O_2 under normal pressure (Lamberdsen C.J., Kough R.H., Cooper D.Y., Emmel G.L., Loeschcke H.H., Schmidt C.F., 1953).

The concentration of oxygen C_{O_2} physically dissolved in the arterial blood in the course of breathing O_2 under pressure $p_{O_2} = 350\,kPa$, as in Table 2.5, is $C \cong 100\% \cdot 2100\,mmHg \cdot 0.003\,cm^3 O_2 \cdot mmHg^{-1} O_2 \cdot 100\,cm^{-3}$ of blood $\cong 6.3\%_v$, and is more than an order of magnitude larger when breathing air under normal pressure: $C \cong 100\% \cdot 91\,mmHg \cdot 0.003\,cm^3 O_2 \cdot mmHg^{-1} O_2 \cdot 100\,cm^{-3}$ of blood $\cong 0.3\%_v$.

Reduced cerebral blood flow produces a 25% compensation of increase in O_2 content physically dissolved in the blood causing the retention of CO_2 resulting from metabolism. This retention is usually not high enough to cause strong symptoms of $hypercapnia$,[43] however, in some cases it can lead to $oxygen\ blackout$ (see Chapter 3). Over time, the increased presence of CO_2 causes the decrease in pH and the blood buffering[44] mechanism cannot counteract this effect, resulting in additional release of O_2 from oxy-hemoglobin into the blood (Figure 2.6). This in turn results in exposure of the brain to increased oxygen partial tensions π_{O_2}, which can form a harmful metabolites.[45] The result can be delayed in time spontaneous *CNSyn* symptoms.

Partial confirmation of the earlier described mechanism could be relatively low hemoglobin saturation, at a level of 89%, in the subclavian vein recorded in the course breathing O_2 under partial pressure of oxygen $p_{O_2} = 350\,kPa$ (Lamberdsen C.J., Kough R.H., Cooper D.Y., Emmel G.L., Loeschcke H.H., Schmidt C.F., 1953).

Suspension of CO_2 removal from cerebral vessels to peripheral venous vessels reduces its concentration and, thus, a decrease in the breathing rate[46] is recorded during the phase preceding *CNSyn*.

A direct impact of CO_2 retention in the peripheral veins on *hypercapnia-induced CNSyn* was not recorded. This retention causes rather regrowth of the ventilation[47] immediately before *CNSyn*.

2.11 SUMMARY

The theory of oxygen window has traditionally been used for planning and assessing the safety of decompression, thus becoming a theory of phenomena that occur in decompression after bounds dive to saturation dives (Behnke A.R., 1967). It has been used for hyperbaric treatment planning (Van Liew H.D., Bishop B., Walder P.D., Rahn H., 1965).

The use of breathing oxygen at the last decompression stations[48] causes the oxygen window to be wide enough to eliminate nitrogen six times faster than in the course of air breathing[49] (Kenny J.E., 1973). Despite the use of oxygen in the last phase of the decompression process, the speed of ascent should be limited due to the kinetics of removal of excess inert gas,[50] which should be removed from the body without formation of the free gas phase. The oxygen window reduces the tendency to form the free gas phase. The toxic effects of oxygen, however, limit the application area of this kind of action. Another disadvantage is the nonlinearity of the window width as a function of oxygen partial pressure in the inhaled breathing medium resulting from the properties of hemoglobin (Vann R.D., 1989) (Figure 2.2).

The concept of *extended oxygen window*[51] helped Professor T. Doboszyński to develop *trimix* (*Tx*) and *nitrox* (*Nx*) tables of continuous decompression after saturation, contributing to a significant success of the Polish hyperbaric science. The concept of such a use of the oxygen window theory is consistent with the findings of other scientists. They, however, have not made any further investigations.[52]

The use of the concept of the oxygen window to explain the phenomena of oxygen toxicity has not so far been reported in the literature, which is why the theories defined in this chapter are not reflected in other reports. Therefore, they should be referred to very carefully.[53] They represent an attempt of theoretical interpretation of the observed phenomena.

It is suggested here that it is a specific exceeded partial oxygen pressure in an inhaled breathing medium that limits the protective effect of the oxygen window and contributes to the risk of developing of *CNSyn*. Therefore, safe exposure to the value exceeding this partial pressure of oxygen is limited in time.

NOTES

1 Biochemical changes associated with the energy conversion taking place in the cells of all living organisms and which underlie all biological phenomena.
2 Forming a pressure difference.
3 Carbon dioxide tension is usually deducted from the value of the oxygen window, but water vapor pressure is not taken into account as it is considered to have constant value.
4 Liquid or bound.
5 *oxygen window* will be about. $\Delta\pi \cong 60$ *mmHg* and will be available for the transport of inert gases.
6 Myoglobin forms a specific oxygen storage used for muscle contraction, especially under the conditions of the so-called oxygen debt; animals that dive have increased amounts of myoglobin.
7 In the form of the complex: MbO_2.
8 This means that at the beginning, an increase in the concentration of O_2 increases the rate of binding by Mb into the complex MbO_2, but above a certain concentration limit, further increase in the concentration of O_2 no longer accelerates MbO_2 complex formation.
9 At this point, the rate of reaction of MbO_2 complex formatio $\dot{v}$ n depends only on the concentration of myoglobin [Mb]: .
10 As will be shown furt $\dot{v} = k \cdot [Mb]$ her, the Michaelis constant is the concentration of oxygen at which oxygen binding reaction rate is half the maximum rate $\dot{v}$ – oxygen affinity P_{50}.
11 $\dot{v} = 0.5 \cdot \dot{v}_{max}$ Myoglobin ability to bind the oxygen.
12 The smaller it is, the higher the affinity is, while a value of this constant precludes low oxygen affinity.
13 Myoglobin-oxygen dissociation model.
14 See Figures 2.2 and 2.3b later in this chapter.
15 substituting into equation $\dfrac{\dot{v}}{\dot{v}_{max}} = \dfrac{[O_2]^n}{K+[O_2]^n}$ value $\dot{v}_{max} = 2 \cdot \dot{v}$, constant K can be calculated, thus, according to the $K = \left(\dfrac{2\dot{v}}{\dot{v}}-1\right) \cdot [O_2]^n = [O_2]^n$ definition of affinity for oxygen, constant K may be repla P_{50} ced with the value of oxygen affinity for hemoglobin $P_{50}^n \leftarrow K$ when oxygen concentration $[O_2]$ is replaced by its partial tension $\pi_{O_2} \leftarrow [O_2]$.
16 Dissociation model of the hemoglobin–oxygen of the complex.
17 In the analysis presented here, the Hill model is sufficiently accurate, but more accurate models of the dissociation of hemoglobin have been considered and are still being developed (Tikuisis P., Gerth W.A., 2003; Lobdell D.D., 1981).
18 Provided that the partial pressure of oxygen in the alveoli is at $p_{O_2} = 100 mmHg$ and the oxygen tension in the capillaries is $\pi_{O_2} = 46 mmHg$ (Figure 2.3).
19 This forms a coordination bond, which means that hemoglobin with oxygen forms a weaker bond than the standard chemical bond.
20 Subject to the Henry law – transfer without formation of a chemical bond.
21 For example, $x_{O_2} = 0.97 \Rightarrow C = 1.39 cm^3 O_2 \cdot g^{-1} O_2 \cdot 150 g Hb \cdot dm^{-3} of blood \cdot 0.97 \cong 202 cm^3 O_2 \cdot dm^{-3}$ of blood.
22 Blood flow in a healthy human being is in the range of $\dot{V} \in [5,6] dm^3$, for the assumed difference in oxygen tension $\Delta\pi_{O_2}$, in accordance with Figure 2.5, oxygen consumption is $13.6 cm^3 O_2 \cdot 100 cm^{-3}$ of blood and hence, global rate of oxygen consumption is approximately

22 $\dot{v}_{O_2} = 136\,cm^3O_2 \cdot dm^{-3}$ of blod $\cdot\,5\,dm^3$ of blood $= 680\,cm^3O_2 \cong 0.7\,dm^3O_2 \cdot min^{-1}$, which represents the light effort as in Table 2.3.

23 Type of hemoglobin with relatively low affinity for oxygen (Kwasiborski P.J., Kowalczyk P., Zieliński J., Przybylski J., Cwetsch A., 2010).

24 $pH = -\log a_{H_3O^+}$, where: activity of $a_{H_3O^+}$ – hydronium ions.

25 Lowering of pH.

26 Easier dissociation of the $Hb(O_2)_n$ complex.

27 Dissociation into carbonate ion is not essential for transport mechanism of carbon dioxide in the blood, since it takes place in the case higher values of $pH > 9$.

28 Creation of the complex CO_2–Hb is accompanied by the emission of hydroniumions (H_3O^+), causing lowered pH in the blood and tissues relative to the lungs.

29 This is discussed in the latter part of the chapter.

30 Here, this syndrome is caused by oxygen toxicity, where syndrome means a group of symptoms which consistently occur together, or a condition characterized by a set of associated symptoms.

31 Which is an equivalent of pressure approximately 0.7 MPa.

32 Partial oxygen pressure was approximately 237 $mmHg$.

33 $x_v = \gamma_{Hb} \cdot C_{Hb} \cdot x_{O_2} \cong 1.39 \cdot 10^{-3}\,dm^3O_2 \cdot g^{-1}Hb \cdot 150\,g\,Hb \cdot dm^{-3}$ of blood $\cdot\,0.996\,$: $\cong 0.208\,dm^3O_2 \cdot dm^{-3}$ of blood, where: y_{Hb} – maximum saturation of Hb with oxygen, C_{Hb} – average content of hemoglobin in the blood.

34 Bradycardia (Lat. *bradycardia*) here is referred to a condition in which the heart rate drops due to the body hyperbaric exposure as compared to the heartbeat in normobaria.

35 During the experiment, divers used inhalators for breathing while sitting in the decompression chamber in comfortable temperature.

36 As regards the analysis conducted here, Hill's model is sufficiently accurate. However, there also exist more detailed models used to describe the total solubility of oxygen in the blood (Tikuisis P., Gerth W.A., 2003); analysis of several other models was given by Lobdell (Lobdell D.D., 1981).

37 Attention should be paid to the decrease in blood flow related to staying under higher pressure and associated with bradycardia phenomena.

38 A cascade of oxygen and other biochemical cycles.

39 Adenosine triphosphate – *ATP*, guanosine triphosphate – *GTP*.

40 Therefore, the drop of the oxygen content should be maintained at the same level at the same diver's effort.

41 In Figure 2.8 and Table 2.5, the difference between points $T1$ and $Z1$ is slightly smaller because it relates to extremely low oxygen consumption (Lamberdsen C.J., Kough R.H., Cooper D.Y., Emmel G.L., Loeschcke, Schmidt C.F., 1953).

42 Saturation of hemoglobin by oxygen is still at 100% in both venous and arterial blood.

43 Hypercapnia is a state of elevated partial pressure of CO_2 in the blood above $p_{CO_2} > 45\,mmHg$, here we have symptoms of CO_2 intoxication.

44 A solution whose value of pH after addition of small amounts of strong acids or alkali, as well as after dilution with water, hardly change.

45 Organic or inorganic product of metabolism.

46 Stimulated by the respiratory center.

47 And thereby exposes the tissues to the increased influence of the oxygen stream physically dissolved in the blood.

48 Outside the area of its toxic effects.

49 For example, when breathing air removal of nitrogen amounts to approximately 0.45 kg in approximately 30 min, but in the course of breathing isobaric oxygen the same amount of nitrogen will be removed from the human body in approximately 5 min.

50 For example Bühlmann, who used in his works the concept of the oxygen window, reduced the rate of diver ascends to the surface to 10 $mH_2O \cdot min^{-1}$. He also introduced a mandatory, one-minute decompression stop at a depth of 3 mH_2O when diving does not require decompression stops. The *American Academy of Underwater Sciences* has provided evidence proving the results of the research done by Bühlmann. It has been shown that the reduction in the rate of ascent and the use of decompression safety stop at zero decompression dives reduces the size of the quiet gas phase minimum six times and the tension of nitrogen in the fast theoretical tissues between 12% and 21% with only a minimal increase in the partial tension of nitrogen in the slower theoretical tissues (Cole B., 1993).

51 Prof. T. Doboszyński assumed, in his investigations, the maximum value for the oxygen window for saturation dives at a level of 150 *mmHg*, calling it extended oxygen window (Kot J., Sicko Z., Doboszynski T., 2015).

52 In the *SEA – LAB II* quoted experiment, the oxygen window was at approximately 150 *mmHg* (Behnke A.R., 1967); the same maximum value of the oxygen window for saturated dives was applied by Prof. T. Doboszyński and was referred to as extended oxygen window.

53 Mainly because they have not been confirmed by our own experiments.

REFERENCES

Åstrand P-O & Rodahl K. 1977. *Textbook of Work Physiology: Physiological Bases of Exercises*. New York: McGraw-Hill, Inc. ISBN 0-07-002406-5.

Behnke AR. 1967. The isobaric (oxygen window) principle of decompression. In *The New Thrust Seaward. Transactions of the Third Annual Conference of the Marine Technology Society Conference, San Diego*. 5–7 June. Washington, DC: Marine Technology Society.

Berg JM, Tymoczko JL & Stryer L. 2013. *Biochemistry*. New York: W.H. Freeman and Company. ISBN: 1-4292-8360-2. ISBN-13: 978-1-4292-8360-1.

Cole B. 1993. *Decompression and Computer Assisted Diving*. Biggin Hill: Dive Information Company. ISBN 0-9520934-0-5.

Corda M, De Rosa MC, Pellegrini MG, Sanna MT, Olianas A, Fais A, Manca L, Masala B, Zappacosta B, Ficarra S, Castagnola M & Giardina B. 2000. Adult and fetal hemoglobin J-Sardegna [α50(CE8)His→Asp]: functional and molecular modelling studies. *Biochem. J.*, 346, 193–199.

Ekeloef NP, Eriksen J & Kancir CB. 2001. Evaluation of two methods to calculate p50 from a single blood sample. *Acta. Anaesthesiol. Scand.*, 45, 550–552. http://doi.org/10.1034/j.1399-6576.2001.045005550.x.

Kenny JE. 1973. *Business of Diving*. Houston: Gulf Publishing Co. ISBN 0-87201-183-6.

Kłos R. 2021. *Ventilation of Normobaric and Hyperbaric Objects*. Boca Raton, FL: CRC Press (Taylor & Francis Group, LLC). ISBN: 978-0-367-67523-3 (hbk). ISBN: 978-0-367-67524-0 (pbk). ISBN: 978-1-003-13164-9 (ebk).

Kot J, Sicko Z & Doboszynski T. 2015. The extended oxygen window concept for programming saturation decompressions using air and Nitrox. *PLoS ONE*, 6, 1–20. http://doi.org/10.1371/journal.pone.0130835.

Kwasiborski PJ, Kowalczyk P, Zieliński J, Przybylski J & Cwetsch A. 2010. Znaczenie powinowactwa hemoglobiny do tlenu w adaptacji do hipoksemii. *Polski Merkuriusz Lekarski.*, 28, 166–260.

Lamberdsen CJ, Kough RH, Cooper DY, Emmel GL, Loeschcke HH & Schmidt CF. 1953. Oxygen toxicity. Effects in man of oxygen inhalation at 1 and 3,5 atmospheres upon blood gas transport, cerebral circulation and cerebral metabolism. *J. Appl. Physiol.*, 9, 471–486.

Lobdell DD. 1981. An invertible simple equation for computation of blood O_2 dissociation relations. *J. Appl. Physiol.: Respirat. Environ. Exercise Physiol.*, 50, 971–973.

Nunn JF. 1993. *Nunn's Applied Respiratory Physiology.* Jordan Hill: Butterworth-Heinemann Ltd. ISBN 0 7506 1336 X.

Przylipiak M & Torbus J. 1981. *Sprzęt i prace nurkowe-poradnik.* Warszawa: Wydawnictwo Ministerstwa Obrony Narodowej. ISBN 83-11-06590-X.

Stryer L. 1997. *Biochemia.* Warszawa: Wydawnictwo Naukowe PWN. ISBN 83-01-12044-4.

Tikuisis P & Gerth WA. 2003. Decompression theory. In *Bennett and Elliott's Physiology and Medicine of Diving,* ed. AO Brubakk & TS Neuman, 419–494. Edinburgh: Saunders.

Van Liew HD, Bishop B, Walder PD & Rahn H. 1965. Effects of compression on composition and absorption of tissue gas pockets. *J. Appl. Physiol.*, 20, 927–933.

Vann RD. 1989. The physiology of NITROX diving. In *Workshop on Enriched Air NITROX Diving,* ed. RW Hamilton, DJ Crosson & AW Hulbert, 85–104. Rockville: National Oceanic and Atmospheric Administration.

West JB & Luks AM. 2016. *West's Respiratory Physiology – The Essentials.* Philadelphia: Wolters Kluwer. ISBN 978-1-4963-1011-8.

3 Oxygen Toxicity

3.1 CENTRAL NERVOUS SYSTEM TOXICITY

Oxygen is necessary to maintain homeostasis[1] in the human body, but under hyperbaric conditions it causes central nervous system toxicity and toxicity in lung tissues, general toxicity in the other tissues, as well as other side effects.

The Polish Navy has maintained, until recently, the rules for fixing exposure time and maximum allowable oxygen partial pressure during a dive (Kłos R., 2000). They corresponded to the applicable provisions of the US Navy in the 1960s – Table 3.1 (Kenny J.E., 1973).

The manual released by the Ministry of National Defense in 1981 distinguishes between routine and exceptional exposures, where the boundary between them runs at a pressure of 0.175 *MPa* (Przylipiak M., Torbus J., 1981). Currently, during dives performed with closed circulation apparatuses where oxygen is used as a breathing medium, the Polish Navy accepts the recommendations by the US Navy (US Navy Diving Manual, 2008; US Navy Diving Manual, 2016). This monograph proposes amendments to some of these recommendations (see Chapter 5).

3.1.1 CENTRAL NERVOUS SYNDROME

Oxygen toxicity impact on the central nervous system is called the *Paul Bert effect*; sometimes the acronym *CNS* for central nervous syndrome is used, but for the purposes of this book, in order to distinguish it from the acronym used to define the central nervous system (*CNS*), *CNSyn* will be used.

TABLE 3.1

Oxygen partial pressures and exposure time limits according to the regulations in force in the Polish Navy.

Maximum oxygen partial pressure	Allowable exposure time
[MPa]	[min]
0.130	240
0.145	150
0.160	110
0.175	75
0.190	45
0.205	25
0.220	10

DOI: 10.1201/9781003309505-4

According to some sources, in the recent years no significant advances in the knowledge of oxygen toxicity *CNSyn* have been made (Bitterman N., 2004). Still, many researchers refer to classical studies carried out during the Second World War by Kenneth Donald that were updated and published with the new findings compiled (Donald K.W., 1992). The most commonly used scenarios of combat exposure to oxygen were developed in the 1920s (Butler F.K., Thalmann E.D., 1984; Butler F.K., Thalmann E.D., 1986a; Butler F.K., Thalmann E.D., 1986b).

The phenomenon of oxygen toxicity of *CNSyn* is difficult to study because of the complex interaction of many factors, which include age, gender, individual predisposition and current mental and physical condition (Donald K.W., 1992). Toxicity is a phenomenon affecting many organs important for maintaining homeostasis. Frequently the biochemical systems theory is used to explain the phenomena of oxygen toxicity, assuming the adverse effects generated by the free radicals[2] and other metabolites that manifest as *CNSyn* symptoms (Torbati D., Church D.F., Keller J.M., Pryor W.A., 1992; Bartosz G., 2008).

During the oxygen exposures, hydrogen peroxide (H_2O_2) was observed in the human blood and its effect was tested on various brain areas (Bitterman N., 2004).

It is suggested that oxygen exposure has an effect on various neuroreceptors,[3] e.g. *GABA*[4] receptors, and that higher oxidized metabolic forms have an effect on enzymes[5] important for the human body, for example acetylcholine[6] (Bitterman N., 2004).

Mentioned among the methods used for reducing the risk of *CNSyn* is the introduction of air breaks that can be applied during oxygen decompression or hyperbaric treatment *HBOT*[7] (Clark J.M., Thom S.R., 2003).

During exposure to oxygen, an initial narrowing of cerebral blood vessels was recorded. This resulted in decreased blood flow to the cerebral cortex,[8] followed by an expansion of blood vessels. The moment at which this relaxation occurs seems to be the *CNSyn* threshold point, beyond which the *CNSyn* symptoms develop until an attack of convulsions.[9] Substances that reduce blood flow through the cerebral cortex can inhibit the development of *CNSyn* symptoms (Bitterman N., 2004).

3.1.2 THE MECHANISM OF CENTRAL NERVOUS SYNDROME

The *CNSyn* phenomenon will be briefly presented here using, as an example, the general biochemical theory of oxygen toxicity.

Biochemical transitions that utilize oxygen are the source of energy for higher forms of life on earth. The energy necessary to sustain life is obtained by the oxidation reactions occurring in cells. Bond energy released during the oxidation of carbohydrates, proteins and fats is stored in portions in phosphate bonds of guanosine-5'-triphosphate (*GTP*) and adenosine triphosphate (*ATP*) (Berg J.M., Tymoczko J.L., Stryer L., 2013). A simplified diagram of obtaining *GTP* and *ATP* in a cell is presented in Figure 3.1.

The major part of *ATP* is obtained from the cycle of the *respiratory chain*.[10] In the respiratory chain there occurs a combustion of hydrogen transported by such enzymes as nicotinamide adenine dinucleotide (*NAD*) from the *Krebs cycle*.[11] Oxygen is delivered to the respiratory chain by *cytochromes*.[12]

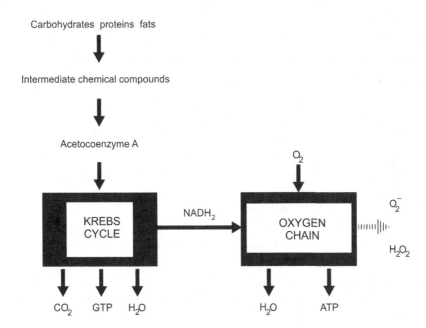

FIGURE 3.1   General diagram of *ATP* and *GTP* production.

Cytochromes contain an iron atom capable of binding and donating oxygen. Then, a change in the valence state of iron takes place and cytochromes change from the oxidized form to the reduced form and vice versa. In a reaction of oxygen with $NADH_2$, water is produced and the reaction energy, which is stored in the phosphate bonds of *ATP*, is released. This is the main method of producing *ATP* that occurs in cell *mitochondria*[13] (Stryer L., 1997; Berg J.M., Tymoczko J.L., Stryer L., 2013).

When oxygen tension in tissues is high, oxygen may enter the respiratory chain in large quantities. Then the biochemical reactions lead to the formation of free radicals and H_2O_2. Responsible for this reaction are *redox enzymes*, which accelerate the *redox reactions*.[14] The resulting free radicals and H_2O_2 are potentially toxic to the cell,[15] but normally they should be deactivated[16] (Bartosz G., 2008; Nowotny F., Samotus B., 1971).

With a significant increase in oxygen tension, the production of toxic compounds rises and the biochemical protection system is not able to deactivate them, resulting in biochemical and physiological changes in the functioning of the body. They manifest themselves as *CNSyn*[17] symptoms. No cases of these symptoms occurring immediately after exposure of the body to oxygen have been recorded.[18] As a result of poisoning, some *CNSyn* symptoms were observed preceding an attack of convulsions. They included the following: restlessness, pale face, trembling lips and eyelids, nausea, cramps, dizziness, lack of coordination, visual and auditory hallucinations, narrowing the field of vision[19] and speech disorders (Table 3.2).

These symptoms are rarely noticeable before the onset of seizures[20] (Harabin A.I., Survanshi S.S., 1993; Brubakk A.O., Neuman T.S., 2003).

TABLE 3.2
Symptoms and signs of oxygen poisoning (Harabin A.L., Survanshi S.S., Homer L.D., 1994; Harabin A.L., Survanshi S.S., Homer L.D., 1995).

The degree of risk	Symptoms	Number of cases
	Nausea	75
	restlessness, dyspnea, insomnia, depression	12
	headache	5
	numbness, burning sensation	13
	dizziness	63
	cramps	335
	auditory disorder	7
	vision disorder	17
	loss of consciousness, speech disorder	16
	convulsions	91
Total		634

Generalized convulsions occur suddenly. The attack begins with the tonic phase usually lasting 30 s, during which the diver loses consciousness and stops breathing. This is followed by the clonic phase of uncoordinated movements of the entire body. The whole attack usually lasts about 2 min. After a long period of breathing oxygen in a hyperbaric chamber, where oxygen can be replaced with air, the allowable period of apnea, without causing harm to the poisoned diver, is about $t \in$ [5, 8] min[21] (Clark J.M., Thom S.R., 2003).

Factors that increase cerebral blood flow, such as immersion, hypothermia, workload, increase of CO_2 concentration and so on, increase the sensitivity to $CNSyn$ symptoms (Vann R.D., 1993). Carbon dioxide can be present in the inhaled breathing medium or come from the so-called dead space.[22] Through the CO_2 tension receptors, the body increases the intensity of ventilation. The increase in ventilation rate is also accompanied by an increased density of inhaled breathing medium, increased breathing resistance and so on.

In norm-baric conditions, the excess of CO_2 could be more effectively eliminated than under hyperbaric conditions. Generally, under increased pressure CO_2 accumulates in the body. Initial hyperventilation leads to $hypocapnia$,[23] observed during exposure to oxygen, resulting in decrease of respiratory action. The defense mechanism that causes blood vessels in the brain to narrow leads to increased concentration of CO_2 in cerebral vessels in relation to peripheral vessels (see Chapter 2). Thus, CO_2 receptors in the initial phase cannot increase ventilation. This defense mechanism, however, has its drawbacks. The increased CO_2 concentration in cerebral vessels increases the concentration of hydronium ions (H_3O^+), which results in hemoglobin rapidly losing oxygen in blood and increasing its tension in plasma.[24] Because of this, brain tissue is exposed to higher oxygen tension π_{O_2}. It seems, however, that the body tries to compensate these effects with the enzymes that deactivate the free radicals produced.

During work, CO_2 emission from tissues to peripheral blood may result in the following: increase in ventilation, vasoconstriction at the periphery, which raises blood pressure, and expansion of cerebral blood vessels. This may increase the cerebral blood flow, potentially causing an increase in the O_2 stream flowing through brain.

3.1.3 ANTIOXIDANTS

As already mentioned, during the increase of oxygen tension π_{O_2} in tissues, oxygen can enter the respiratory chain in increased volumes, resulting in the formation of free radicals and superoxides. It is assumed that under normal circumstances, they are deactivated by antioxidants. One effective antioxidant is melatonin.[25] Its action has been widely reported.

However, the activity of melatonin is spread over time and does not result in shifting the threshold of *CNSyn* occurrence (Swiergosz M.J., Keyser D.O., Koller W.A., 2004). Additional intake of melatonin causes somnolence, so its use is limited only to the time scheduled for rest.

An assessment of effectiveness of antioxidants depends on the test method. For example, a study in which healthy men exposed to hyperbaric oxygen had a diet enriched with *vitamins C* and *E* did not confirm any significant antioxidant capacities of these vitamins (Bader N., Bosy-Westphal A., Koch A., Rimbach G., Weimann A., Poulsen H.E., Müller M.J., 2007). Earlier studies conducted on mice infected with malaria[26] showed significant antioxidant activity of isoascorbic acid,[27] which is *isomer L* of ascorbic acid[28] (Rencricca N.J., Coleman R.M., 1979) – Figure 3.2.

The difference between the effectiveness of the optical[29] isomers of ascorbic acid was tested and proved to be significant. However, it was not taken into account in recent studies (Bader N., Bosy-Westphal A., Koch A., Rimbach G., Weimann A., Poulsen H.E., Müller M.J., 2007).

3.1.4 OXYGEN EXPOSURES

It is understood that O_2 shows no toxic effects on the central nervous system, when partial pressure P_{O_2} is equal to or less than $p_{O_2} \leq 0.1\,MPa$[30] (Betts E.A., 1992) – the rationale for adopting this value is given in Chapter 4. During combat dives, oxygen partial pressure p_{O_2} often goes beyond this value.[31]

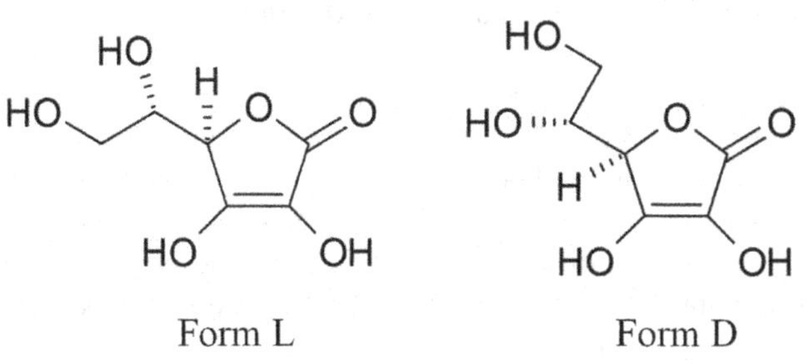

Form L   Form D

FIGURE 3.2 Optical isomers of ascorbic acid.

TABLE 3.3
Partial pressure and time limits for oxygen exposure accepted by the US Navy in the early 1990s (US Navy Diving Manual, 1980).

Oxygen partial pressure	Standard exposures	Exceptional exposures
[MPa]	[min]	[min]
0.10	240	*
0.11	120	*
0.12	80	*
0.13	60	240
0.14	50	180
0.15	40	120
0.16	30	100
0.17	**	80
0.18	**	60
0.19	**	40
0.20	**	30

* times limited only due to L. Smith effect
** exposures prohibited during standard dives

In the 1970s, the US Navy changed the norms of oxygen exposure due to the risk of *CNSyn* symptoms appearing. The changes covered reduction in the partial pressure limits to $p_{O_2} \leq 0.20\,MPa$, division of the exposure type into standard and exceptional, and reduction in allowable time limits for each of the two exposure types. Accepted (until the early 1990s) oxygen exposure time limits during dives with oxygen are presented in Table 3.3 (US Navy Diving Manual, 1980).

In the 1990s, the US Navy changed the allowable oxygen exposure times for *CCR − SCUBA*[32] to allow for more flexible diving operations (US Navy Diving Manual, 2016). In 1991, NOAA[33] changed its rules concerning the partial pressure limits of oxygen during *Nx* exposures[34] (NOAA, 2001; NOAA, 2017). The amendment covered a longer time of stay under the allowable partial pressures of oxygen in *Nx* and determination of the maximum exposure time during the 24 *h* period (Table 3.4).

During deep dives, breathing resistance increases due to an increase in density of breathing medium. The breathing resistance may be accompanied by a rise in accumulation of CO_2 in the diver's body for the reasons described previously. With the increasing distance from the water surface, the possibility of providing assistance to or of a self-rescue of a diver/divers who has/have developed oxygen poisoning becomes more complicated. Therefore, the maximum permitted partial oxygen pressure for deep dives should be reduced.[35]

TABLE 3.4
Exposure time and partial pressure limits of oxygen in the Nx accepted by NOAA (NOAA, 2001; NOAA, 2017).

Standard exposures

Oxygen partial pressure	Exposure time limits		Maximum exposure time in 24 h	
[MPa]	[min]	[h]	[min]	[h]
0.16	45	0.75	150	2.5
0.15	120	2.0	180	3.0
0.14	150	2.5	180	3.0
0.13	180	3.0	210	3.5
0.12	210	3.5	240	4.0
0.11	240	4.0	270	4.5
0.10	300	5.0	300	5.0
0.09	360	6.0	360	6.0
0.08	450	7.5	450	7.5
0.07	570	9.5	570	9.5
0.06	720	12.0	720	12.0

Exceptional exposure

0.20	30	0.50		
0.19	45	0.75		
0.18	60	1.00		
0.17	75	1.25		
0.16	120	2.0		
0.15	150	2.5		
0.14	180	3.0		
0.13	240	4.0		

NOTE:
– if in one of the dives the time of exposure has been reached or has exceeded the limit, the diver must rest for at least 2 h on the surface before next exposure
– if in one or more dives over a period of 24 h the maximum exposure time assumed for 24 h has been reached or exceeded, the diver must rest for at least 12 h on the surface before next exposure (Rutkowski D., 1990)

Similarly, when diving in caves or wrecks, because of limited access to the surface, the rules of deep dives should be followed. The distinction between the standard exposure and the exceptional exposure is related to the conditions of a particular dive. The exceptional exposure can be allowed only for the purpose of saving a human life or other important random events.

3.2 PULMONARY TOXICITY

Oxygen is toxic to the respiratory system. Oxygen toxicity was first observed during prolonged[36] breathing of pure oxygen at atmospheric pressure – and it was called either *Lorrain Smith effect*, from the name of the discoverer, or oxygen pulmonary toxicity (Shilling C.W., 1981; Brubakk A.O., Neuman T.S., 2003). The symptoms of pulmonary toxicity are very similar to *pneumonia*.[37] During dives outside the saturation dive protocol, this effect is minor compared to the *Paul Bert effect*,[38] but it is recommended that it should be monitored during oxygen dives (Shilling C.W., 1981; Brubakk A.O., Neuman T.S., 2003).

3.2.1 UNITS OF PULMONARY TOXICITY

At the end of the 1960s, a pulmonary toxic dose unit of oxygen, known as *UPTD*,[39] *pulmonary oxygen toxicity Q*, was established as an equivalent to one-minute exposure at oxygen partial pressure $p_{O_2} = 0.1\,MPa$ (Harabin A.L., Homer L.D., Weathersby P.K., Flynn E.T., 1987).

3.2.2 REPEX

The most commonly used system of protection against pulmonary toxicity is the method validated during the *Repex* program (Hamilton R.W., Kenyon D.J., Peterson R.E., 1988; Hamilton R.W., Kenyon D.J., Peyerson R.E., Butler G.J., Beers D.M., 1988). However, there are many other models described in the literature (Shykoff B., 2007; Arieli R., 2019). Intensive research in this area is associated with the development of technical diving and medical uses of O_2, but some research is also being carried out in military applications (Arieli R., Yalov A., Goldenshluger A., 2002).

In the Repex system, it is assumed that O_2 begins to be toxic to the lung tissue if it exceeds partial pressure $p_{O_2} > 0.05\,MPa$.[40] To calculate pulmonary oxygen toxicity Q Table 3.5 can be used, which provides minute doses of $\dot{\varrho}$ as a function of oxygen partial pressure p_{O_2}.

For example, breathing pure O_2 for $t = 30\,min$ under pressure $p_{O_2} = 0.15\,MPa$ exposes the diver to *pulmonary toxicity dose Q* of approximately $\varrho = \dot{\varrho} \cdot \tau = 1.77\,UTPD \cdot min^{-1} \cdot 30\,min \cong 53\,UPTD$. For the same purpose a functional dependency can be used (Hamilton R.W., 1989; Shykoff B., 2007):

$$\varrho = t \cdot \left(\frac{p_{O_2} - p_{O_2 max}}{p_{O_2 max}} \right)^{\frac{5}{6}} \tag{3.1}$$

where: p_{O_2} – oxygen partial pressure, t – time, $p_{O_2 max}$ pressure under which pulmonary toxicity symptoms are not observed $p_{O_2 max} = 0.05\,MPa$, $\frac{5}{6}$ – exponent for the best model approximating the experimental data.

The maximum safe doses of pulmonary oxygen toxicity Q depend on exposure time t in Table 3.6, and in Table 3.7 the average values of the vital lung capacity

TABLE 3.5
Values of minute dose of $UTPD \cdot min^{-1}$ of pulmonary oxygen toxicity ϱ as a function of oxygen partial pressure p_{O_2} (Hamilton R.W., 1989).

Oxygen partial pressure p_{O_2}	Minute dose of oxygen pulmonary toxicity ϱ	Oxygen partial pressure p_{O_2}	Minute dose of oxygen pulmonary toxicity ϱ
[MPa]	[$UTPD \cdot min^{-1}$]	[MPa]	[$UTPD \cdot min^{-1}$]
0.05	0.000	0.16	1.92
0.06	0.265	0.17	2.01
0.07	0.490	0.18	2.20
0.08	0.656	0.19	2.34
0.09	0.831	0.20	2.48
1.00	1.00	0.21	2.61
0.11	1.16	0.22	2.74
0.12	1.32	0.23	2.88
0.13	1.47	0.24	3.00
0.14	1.62	0.25	3.14
0.15	1.77		

TABLE 3.6
Allowable dose of pulmonary oxygen toxicity Q during a multiday oxygen exposure.

Exposure time	Allowable daily dose of pulmonary oxygen toxicity Q	Allowable total dose of pulmonary oxygen toxicity Q
[dni]	[$UPTD$]	[$UPTD$]
1	850	850
2	700	1400
3	620	1860
4	525	2100
5	460	2300
6	420	2520
7	380	2660
8	350	2800
9	330	2970
10	310	3100
11	300	3300
12–30	300	

TABLE 3.7
Reduction in the vital capacity of lungs after oxygen exposure and rest time required to eliminate this effect.

Maximum acquired pulmonary dose Q	Reduction of the vital capacity of the lungs	Minimum required time between exposures
[UPTD]	[%]	[h]
615	2	2
825	4	4
1035	6	6
1230	8	8
1425	10	10–12
1815	15	13
2190	20	20

reduction resulting from exposure to oxygen are given together with the average time required to compensate for this effect.

3.3   SOMATIC TOXICITY AND OTHER THREATS

Oxygen is used in a wide range of its partial pressures (Table 3.8).

Aviation and space medicine deals with *hypoxia*. Problems with hypoxia are also important for survival in the high mountains. For divers this phenomenon is important when diving in high-altitude waters. Acclimation before diving and decompression problems in these conditions are caused not only by the lower ambient pressure but also by reduced oxygen concentration C_{O_2}. Sometimes the oxygen content of C_{O_2} is intentionally reduced as in the case of type INERGEN fire extinguishing mixtures type[47] (Table 3.8).

The carcinogenic effect of O_2 accompanies not only diving activities but the very nature of the impact of O_2 on the human body (Bartosz G., 2008). There is a postulated theory which states that O_2 contained in the air and its long-term carcinogenic impact is responsible to a large extent for the effects of aging.[48]

Undoubtedly, hyperoxia and hypoxia are linked to a direct threat of loss of health and life.

3.3.1   CHRONIC OXYGEN TOXICITY

Chronic effects of breathing under increased oxygen partial pressure p_{O_2} are called general O_2 toxicity or oxygen somatic toxicity.[49] One of the described effects is a reversible reduction in the amount of hemoglobin and the number of red blood cells in the blood of saturation divers similar to the increase in their number after a long-term acclimation to hypoxia. These effects in healthy people are reversible but must be taken into account when planning dives or resting after the dive.

TABLE 3.8
The frequently used oxygen partial pressure ranges.

Oxygen partial pressure [MPa]	Specification
0.010	extinguishing gases type *INERGEN®* – it is safe to breath in limited time a mixture of nitrogen, argon, carbon dioxide and oxygen in which oxygen partial pressure falls down to the level of 0.008 *MPa* under condition, that partial pressure of CO_2 will be 0.005 *MPa* (Fire Research, Test, Development and Education Centre, 1993)
0.012	lower limit of safety due to hypoxia
0.016	first symptoms of hypoxia
0.021	normal oxygen partial pressure in atmospheric air
0.035–0.040	typical saturation exposures
0.050	maximum oxygen partial pressure during saturation dives and the beginning of the lung oxygen toxicity[41]
0.10	pure oxygen breathing on surface
0.16	the most commonly accepted upper limit of safety for nitrox dives out of the saturation zone[42] (Hamilton R.W., 1989; NOAA, 2001)
0.20	treatment table *CX*-30 [43] worked out by Comex in 1986 (Comex Marseille, 1986)
0.24	suggestion to use *Nx* 0.4 at a pressure of 0.6 *MPa* in the treatment of diving diseases (Rutkowski D., 1990)
0.25	the upper limit of permitted combat dives with oxygen[44] (Butler F.K., Thalmann E.D., 1986b)
0.28	20 *min* test of oxygen tolerance (NOAA, 2001) and oxygen treatment tables[45]
0.30	suggestion to use *Nx* 0.5 at a pressure of 0.6 *MPa* in the treatment of diving diseases[46] (Rutkowski D., 1990)

Other more dangerous effects are *paresthesia*[50] and avascular necrosis of bone. However, their direct relationship to diving may be difficult to demonstrate clearly (Hamilton R.W., 1989).

3.3.2 THE CHEMISORPTION OF OXYGEN IN THE MIDDLE EAR SYNDROME

An unpleasant effect observed after oxygen dives may be collapse of the tympanic membrane resulting from chemisorption of O_2 from the middle ear cavity.

During change in dive depth, the diver is forced to equalize pressure in the middle ear cavity through the *Eustachian tube*. Multiple change of depth in the course of diving can cause a significant increase in O_2 concentration in the gas space of the middle ear. Even with minor symptoms of cold Eustachian tube, the channel has a reduced patency, and even trained divers[51] are, then, forced to perform strenuous

Valsalva blowout.[52] The oxygen locked in the ear canal undergoes diffusion through the membrane of the oval window of the inner ear where it is absorbed and consumed,[53] resulting in a reduction of pressure in the middle ear and an increase in the outside force exerted on the eardrum.

If this phenomenon happens during a rest period, the diver will wake up with a headache resulting from the strain of the tympanic membrane. Typically, a congestion of the tympanum occurs, i.e. effusions into external and middle ear and increased release of ear wax in the outer ear. It is accompanied by irritating weakening of hearing sensitivity. Usually there is no perforation of the tympanic membrane. This may occur as a result of earlier injuries that reduce the membrane's flexibility and creates scarring.

Another dangerous effect can be an excessive accumulation of dissolved oxygen in *perilymph*,[54] which may be accompanied by its movement resulting in symptoms similar to those of neurological disorders, which occur during a decompression sickness (Farmer J.C., Thomas W.G., 1976). There are recorded cases of irritation of the ear labyrinth causing its increased activity during oxygen decompression or during hyperbaric oxygen treatment. This effect is associated with moving of the *lymph* due to differences in tension of gases dissolved in it. Such cases have been rare during diving operations in Poland. However, there are credible descriptions of such problems in the international literature (Strauss R.H., 1976).

3.3.3 OXYGEN BLINDNESS

Constriction of cerebral blood vessels as physiological effects associated with exposure of brain tissue to high partial pressure of oxygen can be especially dangerous for eyesight.

During hyperbaric oxygen therapeutic procedures[55] cases were recorded of constriction of carotid artery feeding the retina *CRAO*[56] leading up to full blockage[57] and in effect to blindness. Quick restoration of circulation resulted in restored vision (Anderson B., Saltzman H.A., Barbee J.Y., 1965).

It has been also noticed that a disturbing percentage of newborn babies after a long stay in an incubator with an atmosphere enriched with O_2 irreversibly lost their eyesight (Bartosz G., 2008). For this reason, clean air is now used in incubators.

Visual disturbances associated with exposures to high oxygen partial pressure have been studied since the beginning of scientific investigations on diving (Donald K.W., 1947). It has been repeatedly confirmed that oxygen under high pressure has an adverse effect on eyesight (Clark J.M., Thom S.R., 2003).

3.3.4 OXYGEN BLACKOUT

A wide range of the literature deals with the phenomenon of oxygen blackout related to hypoxia during breath-holding dives. The issue of hypoxia occurring while returning to the surface during such dives will not be discussed here. In relation to oxygen combat dives, the term oxygen blackout is usually understood as loss of consciousness by a diver after switching from breathing oxygen to the atmospheric air after the combat mission. One of its reasons may be the already mentioned physiological defense reaction involving the cerebrovascular constriction that results in reduced cerebral blood flow. It is usually accompanied by peripheral vasodilatation, which can lead to *hypothermia* induced by the cooling effect of the aquatic environment.

In hyperbaric conditions, oxygen tension occurs at a higher level in the peripheral blood. Hence, oxygen receptors in the glomerulus will decrease the respiratory rate and reduce the blood flow. Rapid switching to breathing air – which contains less O_2 and is usually combined with the need to get the job done[58] – before the relaxation of cerebral vessels and before increasing cerebral blood flow may cause the phenomenon of hypoxia. In addition, the pressure reduction related to ascent is accompanied by a decrease in the partial pressure of O_2, which reduces respiratory stimulation as in the case of hyperventilation. If this effect occurs in water, the diver runs a risk of choking with water or drowning, but in most cases it only leads to temporary confusion and loss of concentration (NSO, 2020).

Some hypotheses suggest considering the toxic effect of increased CO_2 tension in the cerebral vessels, but in the light of the experiments *in vivo* it appears that the effect of hypercapnia on oxygen blackout is limited. This is because the recorded increase in CO_2 tension is, in normal conditions, tolerated asymptomatically (Lamberdsen C.J., Kough R.H., Cooper D.Y., Emmel G.L., Loeschcke H.H., Schmidt C.F., 1953). Although increasing the tension of CO_2 in the cerebral vessels significantly affects the drop of O_2 in hemoglobin, the effect is not significant in the case of oxygen blackout but rather in the case of the described earlier O_2 toxic effect on brain tissue. It seems that the mechanism of oxygen blackout during the oxygen mission may be associated rather with decreased CO_2 tension in peripheral vessels causing lack of sufficient stimulation for the respiratory center.[59]

3.3.5 OXYGEN BENDS

Animal studies have shown that the occurrence of decompression sickness (*DCS*) symptoms may be caused by compression while using pure oxygen followed by rapid decompression (Donald K.W., 1955; Bennett P.B., Elliott D.H., 1993; Brubakk A.O., Neuman T.S., 2003). However, the course of oxygen-induced *DCS* in this way was much milder than *DCS* induced after the same air exposure (Donald K.W., 1992). *DCS* induced in such a way usually does not require hyperbaric treatment since it undergoes spontaneous compensation in time (Vann R.D., 1989). Although the phenomenon of oxygen bends is possible, it does not pose a significant risk during a normal diving operation with the use of O_2 as breathing medium. It can have a negative impact on the air transport in divers after exposure, for example during a recovery of special group/section.

3.4 OXYGEN TOLERANCE TEST

The procedure of oxygen tolerance test (*OTT*) used in the armed forces is described in Appendix 2. The description covers only the most important information on the purpose and method used to conduct the test.

3.4.1 NATIONAL PRACTICE

In the armed forces when qualifying candidates for diving with an artificial breathing medium, it is recommended that they undergo selection tests in a decompression chamber: a pressure test and an oxygen tolerance test, or both tests simultaneously. A single oxygen tolerance test should be repeated with an interval of at least a week and pressure tests are recommended to be run as often as possible, especially when Polish Navy divers are to be advanced in ranks, and before the periodic medical examination (Szefostwo Ratownictwa Morskiego, 2007).

3.4.2 INTERNATIONAL EXPERIENCE

Discussions of whether to use the *OTT* have been held since the beginning of systematic research into the use of O_2 as breathing medium (Clark J.M., Thom S.R., 2003). The fundamental problem in obtaining a conclusive *OTT* result is the considerable and significant variability in both individuals[60] and groups[61] (Donald K.W., 1947). The same studies found, however, that divers exposed to oxygen poisoning in a decompression chamber and submerged in water showed higher resistance to *CNSyn* while submerged in water. This proves that the test performed in a chamber provides useful information (Donald K.W., 1992). Initially, *OTT* has been used in several countries, but at the beginning of the century it was only the armed forces of Germany and Poland that used it for initial screening.

In 2000, the US Navy conducted an analysis of the results of 6250 *OTT*s in 1976–2000 that involved candidates for combat diving. During this period only six candidates were rejected, hence the benefits to cost ratio was found to be negative and it was recommended that the screening tests in the US Navy[62] be cancelled (Walters K.C., Gould M.T., Bachrach E.A., Butler F. K., 2000). The most important reason, however, for the recommendation to discontinue the screening was the introduction by the US Navy of a new procedure for purging the respiratory volume of the *oxy – CCR SCUBA* with oxygen. As a result of this action, the diver practically breathes $74\%_v O_2/N_2$ which, when enriched with O_2 during a dive, should not exceed $85\%_v O_2/N_2$ (Harabin A.L., Survanshi S.S., Homer L.D., 1994; Walters K.C., Gould M.T., Bachrach E.A., Butler F.K., 2000).

3.4.3 TEST PROCEDURE

A candidate for the *OTT* should have an updated medical certificate and prior to the start of *OTT* should be familiarized with the way the test will be conducted and the rules of behavior inside the hyperbaric chamber.

Before entering the chamber, divers undergoing the *OTT* should answer some questions regarding their general well-being and should be examined by a physician. The scope of examination can be extended but not by elements that could significantly overload the diver on the day of such an examination.[63]

The hyperbaric chamber for running *OTT*s should be prepared appropriately for such tests. The candidate goes through the *OTT* accompanied by an attendant who closely monitors the behavior and who runs the commissioned tests (see Appendix 2).

The tested candidate and attendant enter the chamber, which is then pressurized up to the equivalent of a depth of 18 mH_2O. During the compression process, they breathe the chamber air. After reaching the depth of 18 mH_2O, the tested diver should be again briefly instructed in the use of the oxygen inhaler. Then he puts on the inhaler mask, fits it closely and starts breathing O_2 for 30 min at rest and in recumbence. The period of time from the start of compression to the start of breathing O_2 should be no longer than 15 min. Breathing takes place in the low resistance open system with the discharge of exhaled medium out of the chamber.[64]

During the OTT an attendant, who is breathing the chamber air, monitors the candidate and, at 5 min intervals, carries out measurement of breathing rate, heart rate and blood pressure of the diver undergoing the OTT. The attendant should, at all times, monitor the tested diver for any symptoms of $CNSyn$. If such symptoms are noticed, the subject should make the diver stop breathing oxygen from the inhaler and as soon as possible make him breathe the chamber air.

During the OTT the hyperbaric chamber should be ventilated, preventing concentration of O_2 inside the chamber from rising above $C_{O_2} < 25\%_v$, CO_2 partial pressure of the chamber air should be less than $p_{CO_2} < 1 kPa$.

After 30 min of O_2 breathing, the tested diver removes the mask and for two minutes breathes the chamber air pressurized to the equivalent of 18 mH_2O. Then decompression takes place. The recommended decompression speed should not exceed 10 $mH_2O \cdot min^{-1}$ because of the presence of the attendant. If, in accordance with the Polish Navy air decompression table, decompression stops are not required for the attendant, it is recommended that a 1 min stop be made at a depth of 3 mH_2O, which is considered to be a safety stop.

The first test in a series of OTTs positively completed should be repeated after a minimum of a week-long break. Only the positive results of the second OTT provide a basis for considering the successful completion of OTTs. Such a proceeding is related to the earlier mentioned variability of individuals in O_2 tolerance under hyperbaric conditions (Donald K.W., 1992). Frequently, the $CNSyn$ symptoms detected during the first OTT are not confirmed in the second OTT (Butler F.K., Knafelc M.E., 1986).

If during the OTT[65] the diver convulses the test is terminated, deemed failed and is never repeated. The failed diver can only be authorized for dives with air as breathing medium.

When during the OTT the candidate experienced symptoms of malaise, his/her breathing rate dropped below four breaths per minute or other than seizures symptoms of $CNSyn$ occurred, the OTT should be interrupted and repeated after a minimum gap of one week. Recurrence of symptoms or signs of $CNSyn$ are subject to the same rules as those applicable for cases of seizure occurrence in the tested diver. If the diver during the repeated OTT shows no $CNSyn$ symptoms, this is regarded as successful completion of the first stage of OTT and the second test is performed after a minimum gap of one week. If the third exposure goes without comment, the diver passes OTT. If, however, during the third OTT the diver shows even the slightest sign of $CNSyn$, they should be treated as if they have experienced a seizure.

If the first test is successful, and during the second one malaise symptoms or symptoms of $CNSyn$ other than seizures occur, the OTT may be repeated twice at gaps of at least one week. If, while repeating OTT, symptoms of $CNSyn$ occur, the diver should be treated as if they have had seizures. After each *OTT* the diver should remain in the vicinity of decompression chamber for around one hour.

3.4.4 THE ROLE OF OXYGEN TOLERANCE TEST

Due to shortages of qualified physicians specialized in diving medicine, there is an ongoing discussion concerning regulations applied to diving military units in view of the necessity to lower costs. Given national economic problems and practices in other countries, it seems that resigning from *OTT*s is one of the first permissible steps in this regard.

The *OTT* would have been abandoned a long time ago if there were no common conviction shared by the Polish Navy divers about its usefulness and the rationale behind its implementation. Owing to the positive *OTT* results, the divers have built a sense of security and confidence in their own abilities. A negative result was not used to judge them as unfit for diving. It only imposed limits on their powers to take risks. Additional theoretical and practical training that accompanied *OTT*s consolidated the need for self-control and strengthened the knowledge of *CNSyn*. It must therefore be concluded that the psychological effect was more important than the effects of screening.

It seems that the psychological effect of *OTT* is completely overlooked in the undergoing discussions. Its importance is confirmed by continued employment of psychologists in military units despite the fact that divers are now specially selected professional service members. When it comes to military divers, the psychological effects are marginalized, although a significant reduction in the population of active military divers is recorded, which is balancing on the edge of capability to maintain combat readiness. Probably there are many reasons for this state of affairs. They have one common denominator, i.e. a decrease in the ethos of service, understood here as the standards and values that make up the character of a diving professional, which in turn defines their identity and uniqueness. Not without significance is the effect of lowering the standards of medical service, which also reduces the attractiveness of service in a diving unit. The higher incidence rate of *CNSyn* symptoms recorded in NATO navies prompted some countries to consider reintroducing *OTT*s.[66] That might stop the growing concerns of divers and their defection to other types of service.

Because of the need to rationalize costs, the expenditures incurred by the diving service must be reduced. However, this should not deprive divers of maintaining their feeling of security. It seems that introducing such a pragmatic[67] way of tackling the problems mentioned can only be achieved through organizational changes.

Young and fit candidates, following thorough medical examination during the recruitment, can undergo the selection tests during mandatory training while the medical staff can be used as primary care physician[68] medical diving supervisors and members of recruitment commission. This will allow implementing comprehensive care[69] and will contribute to enhancing the service ethos and strengthening the sense of security among divers. It will also allow for more efficient use of physicians serving in diving units, who also provide a range of psychological services.

3.5   SUMMARY

Oxygen exhibits toxicity to the human body. However, during oxygen dives and in most combat dives it is its neurological form, *CNSyn*, that plays the most important role. This chapter deals with the biochemical view on *CNSyn* associated with the formation of free radicals and forms at a higher oxidation state that can potentially be toxic to the human body. It also includes some considerations concerning the biochemical protection mechanisms against these harmful products, which protect the diver's body from *CNSyn* only to a limited extent.

NOTES

1   The ability of a living organism to retain constant equilibrium of, for example, blood composition or temperature, by proper coordination and regulation of life processes.
2   Group of atoms, generally short-lived, that has unpaired valence electrons – free radical valency.
3   Nervous receptor.
4   Two types of *GABA* receptors binding *γ-amino butyric acid*:
    – receptor $GABA_A$ regulates the influx of chloride ions into the cell, hindering the formation of action potentials responsible for providing information in the nervous system
    – receptor $GABA_B$ regulates the flow of potassium ions and calcium to neutralize the effect of chloride ions and regulate the release of neurotransmitters.
5   The majority of these are proteins, macromolecular chemical compounds regulating the processes of life.
6   Neurotransmitter that converts the electrical signal to a chemical signal at the synapse and that plays a fundamental role in the rapid conduction of nerve signals.
7   Hyperbaric oxygen therapy; sometimes HBO is used as the acronym.
8   Observed as the defensive response to an increase in partial pressure of oxygen (see Chapter 2).
9   Tremor.
10   Electron transport chain.
11   $NADH_2$ – hydrogenated *NAD*.
12   *Cytochromes* are proteins present in cell mitochondria, exhibiting features of biocatalysts involved in the transport of electrons.
13   Mitochondria are surrounded by a membrane structure present in the plasma of most cells with a nucleus, which is where, as a result of cellular respiration, the majority of *ATP* in a cell is produced.
14   Oxidative-reductive.
15   For example, they can cause the inactivation of many enzymes.
16   For example, by vitamin E, catalase, peroxidase, etc.
17   Of these symptoms, convulsions are most often mentioned.
18   The whole biochemical processes described above require time to produce enough radicals to cause DCS, therefore the delay described is indirect evidence of the correctness of the reasoning.
19   Called tunnel vision.
20   However, it is advisable to familiarize divers with them; sometimes trained divers can quickly recognize the beginnings of poisoning, thus preventing severe *CNSyn* symptoms (Harabin A.L., Survanshi S.S., Homer L.D., 1994).

21 Because of good oxygenation of the body.
22 For example, from parts of a diving apparatus with limited gas exchange.
23 Hypocapnia, also hypocarbia, is a state of reduced, below the norm, partial pressure of CO_2 in blood.
24 Together with a decrease in the pH value, the capacity of hemoglobin to bond oxygen is reduced – as in the Bohr effect described earlier.
25 Melatonin is a hormone regulating circadian rhythms, including sleep and wakefulness.
26 Infection results in a significant production of oxidants by protozoa of malaria.
27 Or *E315*, which is a permitted antioxidant used in the food industry to prevent undesirable oxidation processes, extending the shelf life of foods.
28 Which is vitamin C.
29 Frequently observed differences in biochemical activity of optical isomers.
30 NOAA tables adopted this threshold as 0.06 *MPa*.
31 An additional complication is that the oxygen partial pressure during the dive may change in a fairly wide range.
32 Independent closed circuit diving apparatus with oxygen as breathing medium: *Closed Circuit Rebreather (CCR) – Self-Contained Breathing Apparatus (SCUBA)*.
33 National Oceanic and Atmospheric Administration.
34 These changes were preceded by more than 10 years of medical research on dives with nitrox mixtures as breathing medium and *Repex* program (Hamilton R.W., 1989; Hamilton R.W., Kenyon D.J., Peyerson R.E., Butler G.J., Beers D.M., 1988; Hamilton R.W., Kenyon D.J., Peterson R.E., 1988).
35 Often in this type of operation, the stress factor that can significantly increase the cerebral blood flow similarly to the other aforementioned factors is neglected.
36 More than 24 *h*.
37 Symptoms include dry cough, increased breathing resistance, problems with taking full breaths, etc.
38 Paul Bert is considered the first to have described *CNSyn* (Shilling C.W., 1981).
39 Unit of pulmonary toxic dose – *UPTD*, cumulative pulmonary toxic dose – *CPTD*, oxygen tolerance unit – *OTU*.
40 A prolonged stay in the atmosphere, in which the oxygen partial pressure exceeds the $p_{O_2} > 0.05\,MPa$, is limited for the diver's safety.
41 Lorrain Smith effect.
42 The most commonly accepted safety limit for dives using diving apparatus.
43 Use of heliox *Hx* 0.5 at pressure 0.4 *MPa* (Comex Marseille, 1986).
44 Short exposures.
45 Optimal conditions for "washing out" the body from nitrogen.
46 Effective reductions in the size of gas bubbles at the decompression disease.
47 It is worth mentioning that *hypoxia* and *hypercapnia* partially cancel each other out in the atmosphere of INERGEN.
48 For example, the carcinogenic effects of oxygen occurring during long, multiple oxygen exposure were described (Groöger M., Öter S., Simkova V., Bolten M., Koch A., Warninghoff V., Georgieff M., Muth C.M., Speit G., Radermacher P., 2009).
49 Somatic means concerning the body: bodily, physical.
50 Distortion of tactile sensation involving the wrong location of stimuli, and distortion of their experience, as tingling, numbness, prickling sensation, etc.
51 Who do not have problems with equalizing the pressure through swallowing saliva or moving the lower jaw.
52 Blowing out breathing medium from lungs to nose with closed mouth and squeezed nostrils.

53 Physical absorption combined with chemical reaction is chemisorption.

54 The fluid filling the bony labyrinth of the inner ear.

55 Therapeutic procedure often used for general and local tissue oxygenation against anaerobes, after respiration poisoning, to secure the grafts, frostbites, burns, post-radiation damages, osteomyelitis, safeguarding the slow-healing wounds, sudden deafness, etc.

56 Central retinal artery occlusion – *CRAO*.

57 Blockages often concern the main artery branches supplying the retina and can be seen as a partial problem with visual acuity; similar symptoms manifest many cardiovascular diseases, including systemic hypertension, inflammation of the arteries and diseases such as syphilis.

58 For example, associated with the need to remain on the surface of water, going to the shore, leaving the area, counter-action against waves.

59 Any lack of stimulation of the nervous system causes perturbations of conduction of signals, which results in exclusion of control leading to cessation of breathing.

60 Symptoms of *CNSyn* were induced in one diver once or twice a week for a period of 90 days and significant differences in tolerance to increased partial oxygen pressure as a breathing medium were observed (Donald K.W., 1947).

61 Symptoms of *CNSyn* were induced in 36 divers and significant differences in the tolerance time to oxygen partial pressure at .. were observed (Donald K.W., 1947).

62 One has to be aware that to serve in special forces, only candidates with special physical qualities and who are healthy and young are selected; hence, it should be assumed that the threat to the wider population is bigger than that observed during the tests.

63 For example, physical performance tests.

64 Breathing work $W < 0.3\ J \cdot dm^{-3}$ and the maximum instantaneous inhalation resistance $\Delta p < 2.5\ kPa$.

65 The first or the second.

66 Underwater Diving Working Group NATO Standardization Agency Meeting, Istanbul 2011.

67 Understood here as an approach taking into account the logical relationship with the problematic situation.

68 As a substitute for psychologists.

69 Practice shows that the divers in the units may also be subject to other occupational diseases associated not only with diving, for example disorders of the spine associated with running high-speed boats.

REFERENCES

Anderson B, Saltzman HA & Barbee JY. 1965. Retinal vascular and functional response to hyperbaric oxygenation. In *Proceedings of the Third International Conference on Hyperbaric Medicine*, ed. IW Brown & BG Cox, 276–280. Durham: Duke University.

Arieli R, Yalov A & Goldenshluger A. 2002. Modeling pulmonary and CNS O2 toxicity and estimation of parameters for humans. *J. Appl. Physiol.*, 92, 248–256.

Arieli R. 2019. Calculated risk of pulmonary and central nervous system oxygen toxicity: Toxicity index derived from the power equation. *Diving Hyperb. Med.*, 49, 154–160. http://doi.org/10.28920/dbm49.3.

Bader N, Bosy-Westphal A, Koch A, Rimbach G, Weimann A, Poulsen HE & Müller MJ. 2007. Effect of hyperbaric oxygen and vitamin C and E supplementation on biomarkers of oxidative stress in healthy men. *Br. J. Nutr.*, 98, 826–833.

Bartosz G. 2008. *Druga twarz tlenu – wolne rodniki w przyrodzie*. Warszawa: Wydawnictwo Naukowe PWN. ISBN 978-83-01-13847-9.

Bennett PB & Elliott DH. 1993. *The Physiology and Medicine of Diving*. London: W.B. Saunders Company Ltd. ISBN 0-7020-0538-X.

Berg JM, Tymoczko JL & Stryer L. 2013. *Biochemistry*. New York: W.H. Freeman and Company. ISBN: 1-4292-8360-2. ISBN-13: 978-1-4292-8360-1.

Betts EA. 1992. *The Application of Enriched Air Mixtures*. New York: American Nitrox Divers Inc.

Bitterman N. 2004. CNS oxygen toxicity. *Undersea Hyperb. Med. J.*, 31, 63–72.

Brubakk AO & Neuman TS. 2003. *Bennett and Elliott's Physiology and Medicine of Diving*. London: Saunders. ISBN 0-7020-2571-2.

Butler FK & Knafelc ME. 1986. Screening for oxygen intolerance in U.S. Navy divers. *Undersea Biomed. Res.*, 13, 91–98.

Butler FK & Thalmann ED. 1986a. Central Nervous System oxygen toxicity in closed-circuit SCUBA divers II. *Undersea Biomed. Res.*, 13, 193–223.

Butler FK & Thalmann ED. 1986b. *CNS Oxygen Toxicity in Closed-Circuit SCUBA Divers III*. Panama City: USN Experimental Diving Unit. Report No 5-86.

Butler FK & Thalmann ED. 1984. CNS oxygen toxicity in closed-circuit SCUBA diving. In *Proceedings of the Eight Symposium Underwater Physiology*, ed. AJ Bachrach & MM Matzen, 15–30. Bethesda: Undersea Madical Society.

Clark JM & Thom SR. 2003. Oxygen under pressure. In *Bennett and Elliott's Physiology and Medicine of Diving*, ed. AO Brubakk & TS Neuman, 358–418. Edinburgh: Saunders.

Comex. 1986. *Medical Book*. Marseille: Comex.

Donald KW. 1992. *Oxygen and the Diver*. Harley Swan: The SPA Ltd. ISBN 1-85421-176-5.

Donald KW. 1955. Oxygen bends. *J. Appl. Physiol.*, 7, 639–644.

Donald KW. 1947. Oxygen poisoning in man part I. *Br. Med. J.*, 7, 667–672.

Farmer JC & Thomas WG. 1976. Ear and sinus problems in diving. In *Diving Medicine*, ed. RH Strauss. New York: Grune & Stratton, Inc.

Fire Research, Test, Development and Education Centre. 1993. *Test Report: Safety at 8% Oxygen – 15 Minutes*. Copenhagen: Fire research, Test, Development and Education Centre.

Groöger M, Öter S, Simkova V, Bolten M, Koch A, Warninghoff V, Georgieff M, Muth CM, Speit G & Radermacher P. 2009. DNA damage after long-term repetitive hyperbaric oxygen exposure. *J. Appl. Physiol.*, 106, 311–315.

Hamilton RW. 1989. Tolerating exposure to high oxygen levels: Repex and other methods. *Mar. Tech. Soc. J.*, 23, 19–25.

Hamilton RW, Kenyon DJ & Peterson RE. 1988. *REPEX Habitat Diving Procedures: Repetitive Vertical Excursions, Oxygen Limits, and Surfacing Techniques*. Washington, DC: National Oceanic and Atmospheric Administration. Technical Report 88-1B.

Hamilton RW, Kenyon DJ, Peyerson RE, Butler GJ & Beers DM. 1988. *REPEX: Development of Repetitive Excursions, Surfacing Techniques, and Oxygen Procedures for Habitat Diving*. Washington, DC: National Oceanic and Atmospheric Administration. Technical Report 88-1A.

Harabin AI & Survanshi SS. 1993. *A Statistical Analysis of Recent NEDU Single-Depth Human Exposures to 100% Oxygen at Pressure*. Bethesda: Naval Medical Research Institute. NMRI 93-59; AD-A237 488.

Harabin AL, Homer LD, Weathersby PK & Flynn ET. 1987. An analysis of decrements in vital capacity as an index of pulmonary oxygen toxicity. *J. Appl. Physiol.*, 63, 1130–1135.

Harabin AL, Survanshi SS & Homer LD. 1994. *A Model for Predicting Central Nervous System Toxicity from Hyperbaric Oxygen Exposure in Man: Effects of Immersion,*

Exercise, and Old and New Data. Bethesda: Naval Medical Research Institute. NMRI 94-0003; AD-A278 348.

Harabin AL, Survanshi SS & Homer LD. 1995. A model for predicting central nervous system toxicity from hyperbaric oxygen exposure in humans. *Toxicol. Appl. Pharmacol.*, 132, 19–26.

Kenny JE. 1973. *Business of Diving.* Houston: Gulf Publishing Co. ISBN 0-87201-183-6.

Kłos R. 2000. *Aparaty Nurkowe z regeneracją czynnika oddechowego.* Poznań: COOPgraf. ISBN 83-909187-2-2.

Lamberdsen CJ, Kough RH, Cooper DY, Emmel GL, Loeschcke HH & Schmidt CF. 1953. Oxygen toxicity. Effects in man of oxygen inhalation at 1 and 3,5 atmospheres upon blood gas transport, cerebral circulation and cerebral metabolism. *J. Appl. Physiol.*, 9, 471–486.

NOAA. 2001. *NOAA Diving Manual – Diving for Science and Technology.* Flagstaff: Best Publishing Co. ISBN 0-941332-70-5.

NOAA. 2017. *NOAA Diving Manual – Diving for Science and Technology.* Flagstaff: Best Publishing Co. ISBN 9781930536883.

Nowotny F & Samotus B. 1971. *Biochemia ogólna.* Warszawa: PWRiL.

NSO. 2020. *Allied Guide to Diving Medical Disorders.* Brussels: NATO Standardization Office (NSO). ADivP-02 (STANAG 1432).

Przylipiak M & Torbus J. 1981. *Sprzęt i prace nurkowe-poradnik.* Warszawa: Wydawnictwo Ministerstwa Obrony Narodowej. ISBN 83-11-06590-X.

Rencricca NJ & Coleman RM. 1979. *Modulation of Oxygen Toxicity by Select Anti-Melanogenic Compounds.* Lowell: University of Lowell. ADA078239.

Rutkowski D. 1990. *Nitrox Manual.* Key Largo: Hyperbarics International, Inc.

Shilling CW. 1981. *A History of the Development of Decompression Tables.* Bethesda: Undersea Medical Society, Inc.

Shykoff B. 2007. *Performance of Various Models in Predicting Vital Capacity Changes Caused by Breathing High Oxygen Partial Pressures.* Panama City: Navy Experimental Diving Unit. NEDU Report TR 07-13.

Strauss RH. 1976. *Diving Medicine.* New York: Grune & Stratton Inc. ISBN 0-8089-0699-2.

Stryer L. 1997. *Biochemia.* Warszawa: Wydawnictwo Naukowe PWN. ISBN 83-01-12044-4.

Swiergosz MJ, Keyser DO & Koller WA. 2004. *Melatonin Does Not Provide Protection Against Hyperbaric Oxygen (HBO) Induced Seizures.* Silver Spring: Naval Medical Research Institute. NMRC 2004-001.

Szefostwo Ratownictwa Morskiego. 2007. *Tymczasowa instrukcja: Standardowy test ciśnieniowy i tolerancji tlenowej.* Gdynia: Dowództwo Marynarki Wojennej. Załącznik 2 do rozkazu Dowódcy Marynarki Wojennej nr 30/SRM z dn. 02.04.2007.

Torbati D, Church DF, Keller JM & Pryor WA. 1992. Free radical generation in the brain precedes hyperbaric oxygen induced convulsions. *Free Radic. Biol. Med.*, 13, 101–106.

US Navy Diving Manual. 1980. Carson: Best Publishing Co. NAVSEA 0994-LP-001-9010.

US Navy Diving Manual. 2008. Washington, DC: The Direction of Commander Naval Sea Systems Command. 0910-LP-106-0957.

US Navy Diving Manual. 2016. Washington, DC: The Direction of Commander Naval Sea Systems Command. SS521-AG-PRO-010 0910-LP-115-1921.

Vann RD. 1989. The physiology of NITROX diving. In *Workshop on Enriched Air NITROX Diving*, ed. RW Hamilton, DJ Crosson & AW Hulbert. Washington, DC: National Oceanic and Atmospheric Administration.

Vann RD. 1993. Oxygen exposure management. *AquaCorps.*, 7, 54–59.

Walters KC, Gould MT, Bachrach EA & Butler FK. 2000. Screening for oxygen sensitivity in US Navy combat swimmers. *Undersea Hyper. Med.*, 27, 21–26.

4 Basics of Survival Analysis

Survival analysis deals with the statistical modelling of time periods between a selected moment and an expected event. This event is called the result or end point.

Data for survival analysis can also be treated as time to event occurrence, survival time, time to failure, reliability time, duration and so on. The analysis of such data is important in medicine,[1] the social sciences[2] and engineering.[3]

4.1 SYSTEM RESPONSE

The time interval between the starting point and the expected events can be treated as a *random variable*, which is the response of the system – called *survival time*.[4] It should be noted that the starting point must be well defined as there are usually several ways of calculating it.[5]

Time is a continuous variable, therefore survival time T is usually also treated as a continuous random variable. In practice, however, it is observed with a predefined period accuracy[6] and is often expressed on a discrete scale.[7]

Survival analysis is based on statistical inferences regarding the *cumulative distribution F* of survival time T. Most commonly it refers to its simple estimation based on a *single random homogeneous sample*, comparison of survival time T between two attempts[8] and modelling of cumulative distribution F, as a potential function of several *explanatory variables*.

These issues do not differ from the typical inferencing and statistical modelling, but the rationale for special treatment of survival analysis is as follows:

- data for survival analysis are often cut[9]
- standard random variable distributions[10] are often not adequate to model survival cumulative distribution F of survival time T

4.2 CENSORED AND TRUNCATED DATA

When examining an exemplary problematic situation regarding the assessment of the patient's reaction to the applied treatment method, the starting point is usually determined as the patient's inclusion in the group after admission to hospital. Then the time T to the expected event is estimated – such as for example death of the patient.[11]

In practice we have full[12] and cutoff[13] data. For patients who did not die, it is known that their survival time T must be larger than observation time t. This type of data is known as cutoff on the right side – the real value of survival time T is to the right. This data type is the most common among cutoff data and may occur for many reasons.[14]

Another type of incomplete data is the subset cutoff data or left-hand side cutoff data. For section cutoff data survival time T is not exactly known, but it is known

DOI: 10.1201/9781003309505-5

that it is within time interval $T \in (t_1, t_2)$. This means that the expected events did not happen before time t_1, but they occurred before time t_2. Therefore, we can only say that time T is within interval (t_1, t_2). This type of data often occurs in sociological studies carried out for a specific time interval. If for the subset cutoff data the initial value is zero $t_1 = 0$, this type of data is known as a left-hand cut. When the research has to determine the time a specific event in the life of someone, it is always left- or right-hand cut.

An important feature distinguishing the survival analysis from many other mathematical statistical methods is that it can use cutoff data, which may contain important information about the nature of the phenomenon. But this is not the rule, and in this respect one should approach it carefully. As mentioned before, when time t is defined as the closing time for research, than for $T > t$ data will be cut off and for $T > t$ they will be full. If time t is determined in the course of the research, or we decide in advance that the study will be aborted when there is an adequate number of expected events, such a cutting mechanism does not bring any significant data for survival analysis[15] and they cannot be included in the analysis, although survival time T is not, in this case, completely independent of cutoff time t.[16] The data cut mechanism which is significant for survival analysis occurs when the data are functionally connected with survival, e.g. a patient is withdrawn from tests due to effect on survival.[17]

The cutting mechanisms are also important if the patient's response to the applied treatment is negative.

Cutoff data cannot be excluded from survival analysis as they can potentially cause potentially serious deviations when inferences are made,[18] but the presence of certain types of such data means the necessity to use special analytical methods. Occasionally, you can analyze a situation where neither the cutoff data nor the full data can be recorded. Such a situation may occur when you have to make an inference regarding the time to machine failure, was stopped to be used due to expiry[19] of the certificate allowing its further operation or which was sent for a mandatory overhaul. Such data are called *cut off* and require special methods of analysis, which will not be discussed here.

4.3   THE CUMULATIVE DISTRIBUTION OF SURVIVAL TIME

Cumulative distribution F of survival time, with random variable T, should be continuous and positively defined. These conditions, for instance, are met by the cumulative distribution of *gamma function* Γ:[20]

$$\underset{\Gamma(a)=\int_0^\infty x^{a-1} \cdot e^{-x} dx}{\forall} \quad F(t) = \frac{f^a}{\Gamma(a)} \cdot t^{a-1} \cdot exp(-f \cdot t) \qquad (4.1)$$

where: $F(t)$ – cumulative distribution, $L(t)$ – density of probability distribution, f – frequency, t – time.

Distributions frequently used in survival analysis are collected in Tables 4.1 and 4.2.

The most popular distribution in modelling dependencies in survival analysis is the *Weibull distribution*, for which probability density $L(t)$ can be written as:

TABLE 4.1

Generalized gamma distribution.

$$\forall_{x,a,b,c>0} L = \frac{a}{b^{c/a} \cdot \Gamma\left(\frac{c}{a}\right)} \cdot X^{c-1} \cdot \exp\left(-\frac{X^a}{b}\right)$$

Parameter		Distribution	Density of distribution $L(X)$
a	C		
1	1	Exponential	$\forall_{x,b>0} L = \frac{1}{b} \cdot \exp\left(-\frac{X}{b}\right)$
1	c	Gamma	$\forall_{x,b,c>0} L = \frac{1}{b^c \cdot \Gamma(c)} \cdot X^{c-1} \cdot \exp\left(-\frac{X}{b}\right)$
2	2	Rayleigh	$\forall_{x,b>0} L = \frac{2}{b} \cdot X \cdot \exp\left(-\frac{X^2}{b}\right)$
a	a	Weibull	$\forall_{x,a,b>0} L = \frac{a}{b^a} \cdot X^{a-1} \cdot \exp\left[-\left(\frac{X}{b}\right)^a\right]$
2	3	Maxwell	$\forall_{x,b>0} L = \frac{4}{\sqrt{\pi} \cdot b^{3/4}} \cdot X^2 \cdot \exp\left(-\frac{X^2}{b}\right)$

$$\Gamma(a) = \int_0^{\infty} X^{a-1} \cdot e^{-X} dX ; b = \frac{1}{f}$$

$$L(t) = a \cdot f^a \cdot t^a \cdot t^{a-1} \cdot \exp(-f \cdot t)^a \qquad (4.2)$$

where: a – constant.

The parameter a in equation (4.2) is responsible for the shape and frequency f for the scale of the probability density for the Weibull distribution (Figure 4.1).

Average $\bar{x}$ for the Weibull distribution is:

$$\bar{x} = \frac{1}{f} \cdot \Gamma\left(1 + \frac{1}{a}\right) \qquad (4.3)$$

and the variance σ^2:

$$\sigma^2 = \frac{1}{f^2} \cdot \left[\Gamma\left(1 + \frac{2}{a}\right) - \Gamma^2\left(1 + \frac{1}{a}\right)\right] \qquad (4.4)$$

For $a = 1$, the Weibull distribution density L takes the form of an exponential function with an average: $\bar{x} = \frac{1}{f}$.

TABLE 4.2
Some statistical distributions used in the survival analysis (Evans M., Hastings N., Peacock B., 1993).

Distribution	Cumulative distribution $F(X)$	Density $L(X)$	Average $\bar{X}$	Variance σ^2	Survival function $S(X) = 1 - F(X)$	Hazard function $h(X) = \dfrac{L(X)}{S(X)}$	Cumulative hazard function $H(X) = \int_0^t h(X) \cdot dt$
Exponential	$1 - exp\left(-\dfrac{X}{b}\right)$	$\dfrac{1}{b} \cdot exp\left(-\dfrac{X}{b}\right)$	b	b^2	$exp\left(-\dfrac{X}{b}\right)$	$\dfrac{1}{b}$	$\dfrac{X}{b}$
Logistical	$1 - \dfrac{1}{1 + exp\left(-\dfrac{X-a}{b}\right)}$	$\dfrac{exp\left(\dfrac{X-a}{b}\right)}{b \cdot \left[1 - exp\left(-\dfrac{X-a}{b}\right)\right]^2}$	a	$\dfrac{\pi^2 \cdot b^2}{3}$	$\dfrac{1}{1 + exp\left(\dfrac{X-a}{b}\right)}$	$\dfrac{1}{1 + exp\left(-\dfrac{X-a}{b}\right)}$	$log\left[1 + exp\left(\dfrac{X-a}{b}\right)\right]$
Rayleigh	$1 - exp\left(-\dfrac{X^2}{2 \cdot b^2}\right)$	$\dfrac{X}{b^2} \cdot exp\left(-\dfrac{X^2}{2 \cdot b^2}\right)$	$b \cdot \sqrt{\dfrac{\pi}{2}}$	$b^2 \cdot \left(2 - \dfrac{\pi}{2}\right)$	$exp\left(-\dfrac{X^2}{2 \cdot b^2}\right)$	$\dfrac{X}{b^2}$	$\dfrac{X^2}{2 \cdot b^2}$
Weibull	$1 - exp\left[-\left(\dfrac{X}{b}\right)^a\right]$	$\dfrac{a \cdot X^{a-1}}{b^a} \cdot exp\left[-\left(\dfrac{X}{b}\right)^a\right]$	$b \cdot \Gamma\left(\dfrac{a+1}{a}\right)$	$b^2 \cdot \left[\Gamma\left(\dfrac{a+1}{a}\right) - \Gamma^2\left(\dfrac{a+1}{a}\right)\right]$	$exp\left[-\left(\dfrac{X}{b}\right)^a\right]$	$\dfrac{a \cdot X^{a-1}}{b^a}$	$\left(\dfrac{X}{b}\right)^a$

$$\Gamma(a) = \int_0^\infty X^{a-1} \cdot e^{-X} \, dX; \quad b = \dfrac{1}{f}$$

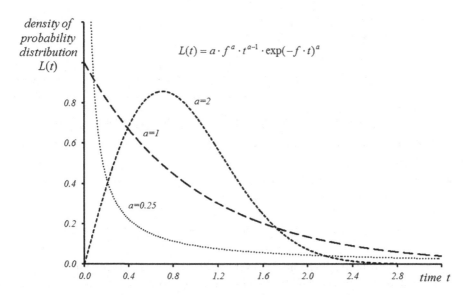

FIGURE 4.1 Density of probability distribution $L(t)$ for the Weibull distribution for different values of parameter a and frequency $f = 1$.

4.4 EXPONENTIAL DISTRIBUTION

For binominal distribution, the probability of the event P with the absence of $n = 0$ undesirable case,[21] with events[22] N, can be written as:[23] $\forall_{n=0} P(n = 0, N) = (1 - R)^N$, where R is the probability of occurrence of an adverse event. In order to express the probability as a function of time $P(n = 0,N) = f(t)$ period of observation t can be divided into a number of sections Δt. Assuming that the value of risk R[24] does not change $R(t) = idem$ for the entire duration of observation t, then with tending to infinity $N \to \infty$ of sections Δt, the probability $P(n = 0,N)$ will be equal to zero:[25] $\forall_{\rho \in (0;1]} \lim_{\Delta t \to 0} P(n = 0, N) = \lim_{N \to \infty} (1 - R)^n \equiv 0$. Assuming that the risk R is constant $R(t) = idem$, the frequency f of occurrence of adverse events is also constant $f = idem$. According to the frequency definition of probability and the earlier assumption that the risk R does not change $R = idem$, one can write (Kłos R., 2007):

$$R = idem = \frac{ES_N}{N} \to R = \frac{f \cdot t}{N} = f \cdot \Delta t \tag{4.5}$$

where: R – risk, ES_N – value of expected E average number of events S_N at population number N, Δt – time section representing the accuracy of counting the time elapsed, N – population number.

Discrete dependent variable ES_N of the expected average number of events n in discrete function of the independent variable of number of observations N can be replaced with continuous dependent variable of risk R as a function of the independent variable of number N of equal sections Δt, where the frequency f will be here a factor of proportionality according to equation (4.5):

$$\frac{ES_N}{N} = f \cdot \frac{t}{N} \to f \equiv \frac{ES_N}{t} \Rightarrow \forall_{\Delta t = \frac{t}{N} = const} \quad R = f \cdot \frac{t}{N} \tag{4.6}$$

Substituting equation (4.6) defining the risk R to the equation: $\forall_{\rho \in (0;1]} \lim_{\Delta t \to 0} P$ $(n = 0, N) = \lim_{N \to \infty} (1 - R)^N \equiv 0$ and using the definition of the exponential function $\forall_{k \in \mathbb{N}} \lim_{k \to \infty} \left(1 + \frac{-a}{k}\right)^k \equiv \exp(-a)$, one can get:

$$\forall_{\rho \in (0;1]} \lim_{\Delta t \to 0} P(n = 0, N) = \lim_{N \to \infty} \left(1 - f \cdot \frac{t}{N}\right)^N \equiv \exp(-f \cdot t) \tag{4.7}$$

where: P – probability, n – number of events in the sample of N cardinality.

The cumulative distribution function $F(t, f)$ for the probability (4.7) can be written by using the definition of inverse probability:

$$\forall_{f>0} F(t, f) = P(0 \le T \le t \,|\, f) = 1 - \exp(-f \cdot t) \tag{4.8}$$

and the density L of the exponential distribution can be found by differentiating the cumulative distribution F in equation (4.8): $\forall_{f>0} L(t, f) = \frac{dF(t,f)}{dt} = f \cdot \exp(-f \cdot t)$. Integrating the density of L in the range from zero to infinity, one can show that the exponential distribution is normalized.[26]

4.5   THE WEIBULL DISTRIBUTION

The probability of a survival additional time Δt, when up to now the life time is t, is the conditional probability:

$$P(\Delta t \,|\, t) = \frac{P(\Delta t \cap t)}{P(t)} \tag{4.9}$$

The numerator of equation (4.9) is the probability of survival of cumulative time $t + \Delta t$; therefore, equation (4.9) can be rewritten into the form:

$$P(\Delta t \,|\, t) = \frac{P(\Delta t + t)}{P(t)} = \frac{exp\left[-f(\Delta t + t)\right]}{exp(-f \cdot t)} = exp(-f \cdot \Delta t) \tag{4.10}$$

From equation (4.10) we can see that the exponential distribution conditional probability of survival of extra time Δt[27] is not a function of the current survival time t: $P(\Delta t | t) \ne f(t)$. This is referred to as the independence of the exponential distribution from current age:[28] $\forall_{P=\exp(-f \cdot \Delta t)} P(\Delta t + t) = P(t) \cdot P(\Delta t)$, but this property is in contradiction with experience.[29]

Modifying this approach by introducing the formula for the density of the probability distribution L, the properties of the exponential distribution $\forall_{f>0} L(t,f) = f \cdot \exp(-f \cdot t)$ can be improved, assuming that the probability of an event for a *single Bernoulli trial*[30] is proportional to the duration of the trial Δt and is independent from the number of the trial (Figure 4.1). Assuming that the probability of an event in an i – trial can be written as: $\forall_{f_i < f_{i+1}; i>1} P_i = f_i \cdot \Delta t$, where frequency f_i of event occurrence is increased $f_i < f_{i+1}$ with the aging of the organism or device. Thus, assuming the independence of trials, probability of $n = 0$ adverse events in N trials will be reflected by the ratio:

$$\forall_{n=0} P(n = 0, N) = \prod_{i=1}^{N}(1 - f_i \cdot \Delta t) \tag{4.11}$$

As before, you can calculate the limit for $N \to \infty \Leftrightarrow \Delta t \to 0$. It is convenient to do so after taking the logarithm of expression (4.11): $\ln P(n=0,N) = \sum_{i=1}^{N} \ln(1 - f_i \cdot \Delta t)$, where the probability can be adopted approximately as follows: $\forall_{\Delta t \to 0} \ln P(n=0,N) \cong -\sum_{i=1}^{N} f_i \cdot \Delta t$. For infinitesimally small Δt we can calculate the limit: $-\lim_{\substack{N \to \infty \\ \Delta t \to 0}} \sum_{i=1}^{N} f_i \cdot \Delta t = -\int_0^t f(t) dt \equiv -\Lambda(t)$. We can therefore write that $\ln P(n=0,N) = -\int_0^t f(t) dt$, and then converting, we can obtain dependence of the probability $P(n = 0, N)$ of the absence of adverse events as a function of time t: $\ln P(n=0,N) = -\int_0^t f(t) dt$. Thus, similarly to equation (4.8) the cumulative distribution $F(t)$ can be written using inverse probability definition: $F(t,f) = 1 - \exp\left[-\int_0^t f(t) dt\right]$. Density L of distribution in this case will be:

$$\forall_{f>0} L(t,f) = \frac{dF(t)}{dt} = f(t) \cdot exp\left[-\int_0^t f(t) dt\right] \equiv f(t) \cdot exp\left[-\Lambda(t)\right] \tag{4.12}$$

In particular, for the density of probability distribution L in equation (4.12), when the frequency $f = \frac{n}{N}$ of the occurrence of the adverse event is not a function of time $f \neq f(t)$, equation (4.12) expresses the density for the *exponential distribution*. For $\Lambda(t) \triangleq t^a$, equation (4.12) expresses the density of the Weibull distribution:

$$\forall_{f>0} L(t,f) = f(t) \cdot exp\left[-t^a\right] \tag{4.13}$$

4.6   THE DENSITY DISTRIBUTION OF SURVIVAL TIME

For the continuous random variable T, i.e. the survival time, from the probability density function L one can determine the distribution of the probability of time T occurrence within the interval (t_1, t_2): $P\left(t_1 \leq T \leq t_2\right) = \int_{t_1}^{t_2} L(t)\,dt$. Cumulative distribution F of time T is given by the formula:

$$F(t) = P(T \leq t) = \int_0^t L(t)\,dt \qquad (4.14)$$

In survival analysis it is often preferred to use three alternative functions of probability defining distribution of the random variable T: *survival function S(t)*, *hazard function h(t)* and *cumulative hazard function H(t)*.

4.7   THE SURVIVAL FUNCTION

Survival function $S(t)$ defines the probability of survival[31] for more than a certain average time t:[32]

$$\forall_{t \geq 0}\ S(t) \equiv P(T > t) = 1 - F(t) \qquad (4.15)$$

where: $S(t)$ – survival function.

Survival function[33] $S(t)$ is a nongrowing continuous function, for which $(0) = 1$. For the Weibull distribution the survival function $S(t)$ is given by the equation: $S(t) = \exp[-(f \cdot t)^a]$. In Figure 4.2, the survival function $S(t)$ is shown for a

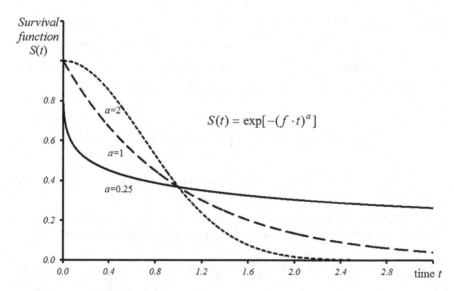

FIGURE 4.2   Survival function $S(t)$ for the Weibull distribution for different values of parameter $a = \{0.25;\ 1;\ 2\}$ and frequency $f = 1$.

Weibull distribution for different values of the parameter $a = \{0.25; 1; 2\}$ and frequency $f = 1$.

The expected value of survival time ET is associated with survival function $S(t)$ by the following formula: $ET = \int_0^\infty S(t)\,dt$, hence this value is represented by the field under the survival function $S(t)$.

4.8   THE HAZARD FUNCTION

Following the definition of conditional probability, we can write:

$$P\left(T \le t + \Delta t \,|\, T > t\right) \equiv \frac{P\left(t < T \le t + \Delta t\right)}{P\left(T > t\right)} = \frac{P\left(T > t \cap T \le t + \Delta t\right)}{P\left(T > t\right)} = \frac{L(t) \cdot \Delta t}{S(t)} \quad (4.16)$$

$$= h(t) \cdot \Delta t$$

where: $h(t)$ – hazard function.

In equation (4.16) the function $h(t)$ is defined as the *hazard function*: $h(t) = \frac{L(t)}{S(t)}$.

From equation (4.16) the hazard function $h(t)$ can be shown as the limit of the conditional probability per time unit:

$$h(t) = \lim_{\Delta t \to 0} \frac{P\left(T \le t + \Delta t \,|\, T > t\right)}{\Delta t} \quad (4.17)$$

Hazard function $h(t)$ represents the probability of survival time T to be in the vicinity of the selected time t, but it will not occur before that time.[34] It describes the intensity of failures for the selected time[35] t. Hazard function $h(t)$ gives the value of probability per unit of time (4.17), therefore, a case[36] in which its value is greater than 1 may occur. For the Weibull distribution, this function is expressed by the equation: $h(t) = f \cdot a \cdot (f \cdot t)^{a-1}$.

Figure 4.3 shows the selected shapes of the hazard function $h(t)$ for the Weibull distribution for different values of the parameter a for frequency $f = 1$.

4.9   THE CUMULATIVE HAZARD FUNCTION

The cumulative hazard function $H(t)$ can be defined as an integral of the hazard function $h(t)$:

$$H(t) = \int_0^t h(t)\,dt \quad (4.18)$$

where: $H(t)$ – cumulative hazard function.

For the Weibull distribution, it is expressed by the equation: $H(t) = (f \cdot t)^a$.

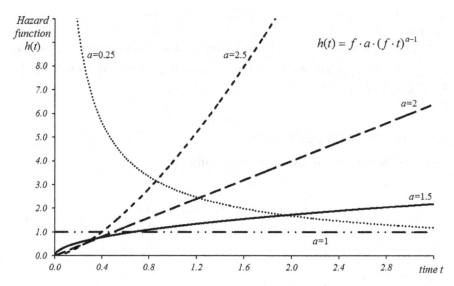

FIGURE 4.3 Hazard function $h(t)$ for the Weibull distribution for different values of parameter a with frequency $f = 1$.

4.10 THE FUNCTION OF RISK

In engineering the hazard function $h(t)$ is sometimes called a *risk function* $R(t)$ or the *failure intensity* $\lambda(t)$ and is defined as the quotient of work time probability density of component $L(t)$ at point t by the probability, for which operating time of component is at least equal t:

$$\forall_{t \geq 0; S(t) \geq 0; \frac{dF}{dt} = \dot{F}(t) \equiv L(t)} h(t) \equiv R(t) \overset{\text{def}}{\equiv} \frac{L(t)}{S(t)} \equiv \frac{\dot{F}(t)}{S(t)} \qquad (4.19)$$

where: $R(t)$ – risk function.

For discrete distribution of probability of intensity of failure $R(t)$, we can write the equation: $\exists_{P_k = P(k \leftarrow X)} R(t) = \frac{P_k}{\sum_{i=k}^{\infty} P_i}$. Differentiating the survival function

$\forall_{t \geq 0} S(t) = P(T > t) = 1 - F(t)$ an interesting relationship can be shown:

$\dot{F}(t) = \frac{d[1 - S(t)]}{dt} = -\dot{S}(t)$, which according to (4.19) gives: $\forall_{S(t) > 0} R(t) = -\frac{\dot{S}(t)}{S(t)} = -$

$\frac{1}{S(t)} \cdot \frac{dS(t)}{dt} \Rightarrow \int_0^t R(t)dt = -\int_0^{S(t)} \frac{dS(t)}{S(t)} = -\ln|S(t)| = -\ln S(t)$. It results in the following

relations:

$$\forall_{S(t) > 0} S(t) = exp\left[-\int_0^t R(t)dt\right] = exp\left[-H(t)\right] \qquad (4.20)$$

Equation (4.20) is called the *Wiener relation*. In engineering, cumulative hazard function $H(t)$ is referred to as distribution *function of safety unreliability*.

4.11 INTERFUNCTION DEPENDENCIES

Summarizing the review of the basic functions used in survival analysis, we can conclude that it is sufficient to provide one of them to describe the other since they are related to each other by the dependencies collected in Table 4.3.

In survival analysis most often survival function $S(t)$ and hazard function $h(t)$[37] are used. In practice it is known that a reliable estimate of the specific functions of survival analysis can be done if the sample size is larger than $N = 30$, otherwise the estimation results are burdened. Despite the convergence of different formulas to calculate the characteristic parameters considered here as timelines, there are differences of interpretation between the theory of reliability, safety analysis, and other applications of survival analysis.[38]

4.12 HAZARD

Risk function $R(t)$ can represent the probability of occurrence of *CNSyn* or *DCS* symptoms in function of time t. The formulas $\forall_{t \geq 0}\, S(t) = P(T > t) = 1 - F(t)$ and $\forall_{S(t) > 0}\, S(t) = exp\left[-\int_0^t R(t)\, dt \right]$ can be used to count the cumulative distribution

---

TABLE 4.3

Useful dependencies for more frequent functions in survival analysis.

$$S(t) = 1 - F(t) = \int_t^\infty L(t)\, dt$$

$$L(t) = -\frac{d}{dt} \cdot S(t)$$

$$R(t) \equiv h(t) = -\frac{d}{dt} \cdot \ln S(t)$$

$$H(t) = -\ln S(t) = \int_0^t R(t) \cdot dt \equiv \int_0^t h(t) \cdot dt$$

$$S(t) = exp\left[-H(t) \right] = exp\left[-\int_0^t h(t) \cdot dt \right]$$

F – cumulative distribution	L – probability density
h – hazard function	R – risk function
H – cumulated hazard function	S – survival function

$F(t)$ of the probability of *DCS* or *CNSyn* symptoms occurrence in function of time expressed by the value of risk function $R(t)$:

$$\forall_{\xi(t)\triangleq F(t)} F(t) \equiv 1 - S(t) = 1 - exp\left[-\int_0^t R(t) \cdot dt\right] \tag{4.21}$$

where: $\xi(t)$ – function of *DCS* or *CNSyn* risk identically equal to cumulative distribution of survival time $F(t)$.

Integral of risk function $R(t)$ from time $t = 0$ to t defines the integral of risk occurrence of *CNSyn* or *DCS* during this period of time. Therefore the value of the risk function $R(t)$ from equation (4.21) determines the hazard function $\xi(t)$ of the onset of symptoms of *DCS* or *CNSyn*. The values of risk function $R(t)$ can be determined by matching them to the experimental data. The limits of integration can be also extended to several hours after the dive.

Using survival analysis to mathematical modelling of the risk function $R(t)$ and of hazard function $\xi(t)$ of the onset of *DCS* or *CNSyn* symptoms, these two issues must be distinguished. The risk $R(t)$ is here identified with the probability of *DCS* or *CNSyn*, and the hazard of *DCS* or *CNSyn* is identified with completing the survival function $S(t)$, which is a cumulative distribution of survival: $\xi(t) = 1 - S(t)$. Hazard $\xi(t)$ is the probability of *DCS* or *CNSyn* occurrence provided the level of risk $R(t)$ of *DCS* or *CNSyn* occurrence is accepted.

4.13   RISK OF DECOMPRESSION SICKNESS

Decompression sickness (*DCS*) can take the classic model of tissue supersaturation as a value of risk function $R(t)$. The risk of *DCS* can be calculated for a set of theoretical tissues. Denoting $R_i(t)$ as a function of risk for the ith theoretical tissue, the theoretical survival function $\forall_{S(t)>0} S(t) = exp\left[-\int_0^t R(t)dt\right]$ can be written as:[39]

$$\forall_{R(t)=\sum_i R_i(t)} S(t) \equiv \prod_i S_i(t) = \prod_i exp\left[-\int_0^t R_i(t) \cdot dt\right] = exp\left[-\int_0^t \sum_i R_i(t) \cdot dt\right] \tag{4.22}$$

A simple model of the risk function $R(t)$ is shown here as an example of matching experimental data using the method of maximum likelihood.

According to the practice with the lapse of time from the end of the decompression, the likelihood of *DCS* occurrence tends to zero. As an algebraic model of *DCS* risk, the decreasing exponential function of time can be accepted: $R(t) = exp(-c \cdot t)$.

There is a moment of time $t = \tau$ after which there will no longer be *DCS* symptoms related to the earlier hyperbaric exposure. As regards the time τ there are different opinions, but most often it is assumed that it is a period not longer than three days. In the adopted mathematical model we can assume that $\xi(t = \tau) \equiv \xi(t \to \infty)$. Using this assumption and relation (4.20), the survival function $S(t)$ for the risk $R(t) = \exp(- c \cdot t)$ can be written as: $S(t) = exp\left[-\int_0^\infty \exp(-c \cdot t) \cdot dt \right]$. The value of the integral can be calculated[40] as: $\int_0^\infty e^{-c \cdot t} \cdot dt = -\frac{1}{c} \cdot e^{-c \cdot t} \big|_0^\infty = \frac{1}{c}$. Hence, the value of the survival function is: $S = \exp(-c^{-1})$ and the hazard of $\xi = 1 - \exp(-c^{-1})$.

DCS or CNSyn occurrence should be treated as an independent event, having cardinality of n for a selected random sample having cardinality of N. Similarly, absence of DCS or CNSyn symptoms is regarded as events independent of each other having cardinality of $N - n$ for the selected random sample having cardinality of N. Thus, the probability function Φ defining cumulative probability of an event can be written as the quotient of probabilities of partial events: $\Phi = \xi^n \cdot S^{N-n}$. Because the probabilities are numbers smaller than 1, their quotient is generally much smaller than 1. Therefore, it is more convenient to use logarithmic probability function $\psi \equiv \ln \Phi = n \cdot \ln(1 - S) + (N - n)$. On the basis of experimentally performed dives, the exact value of the parameter c cannot be calculated because the general population can never be examined, only a sample selected from it at random can. However, its most reliable value can be calculated. To this end the maximum of probability function Φ^{41} must be found. The necessary condition for the extreme of the function to exist is zeroing out its first derivative $\frac{d}{dS} \Phi \equiv 0$. This condition is equivalent to zeroing

out the logarithmic probability function $\frac{d}{dS} \Psi \equiv 0 = -\frac{n}{1-S} + \frac{N-n}{S} \Rightarrow S = \frac{N-n}{N}$. For $S = \exp(- c^{-1})$, $\exp\left(-c^{-1}\right) = \frac{N-n}{N} \Rightarrow c = \left[\ln\left(\frac{N}{N-n}\right) \right]^{-1}$ can be written.

If these criteria were applied to an experiment in which 100 dives were made and where 20 cases of DCS were recorded and in 80 cases no DCS occurred, it can be written that hazard ξ is[42] $\xi = 1 - \exp(-c^{-1}) = 0.2 \to c \cong 4.48$ and hence the risk function $R(t)$ will be (Wienke B.R., 2003) $R(t) \cong \exp(-4.48 \cdot t)$.

Some of the first attempts to apply survival analysis to evaluating the DCS risk were made by Hills, who showed that the survival function $S(\rho) = f(H)$ dependent on risk $R = \frac{n}{N}$ in the coordinates of depth H for eight-hour saturation has Weibull distribution. He did this by showing a good linear dependence of the logarithm of survival function $\ln S$ on the logarithm of shifted saturation depth H (Hills B.A., 1997):

$\forall_{H>0} S\left(R = \frac{n}{N}\right) = \exp\left[-\left(\frac{H-a}{b}\right)^c \right] \to \ln[-\ln S(\rho)] = c \cdot \ln\left(\frac{H-a}{b}\right)$, where: a, b, c – constants of proportionality, H – depth for an eight-hour-long exposure, n – number of cases of DCS, N – number of replications for depth H.

Survival analysis is one of the most promising methods for predicting risk of DCS, which is bridging the statistical and deterministic methods (Brubakk A.O., Neuman T.S., 2003).

4.14   RISK OF CENTRAL NERVOUS SYNDROME

Survival analysis has been also applied to predict the *central nervous syndrome* (*CNSyn*) risk. One of the proposed algebraic models of the hazard function $h\left(t, p_{O_2}\right)$ is the equation[43] (Harabin A.L., Survanshi S.S., Homer L.D., 1994)

$$\forall_{p_{O_2} \geq p_g} \; R\left(t, p_{O_2}\right) = a_0 \cdot a_2 \cdot \left(p_{O_2} - p_g\right)^{a_1} \cdot t^{a_2 - 1} \tag{4.23}$$

where: $R\left(t, p_{O_2}\right)$ – risk function of *CNSyn* occurrence, p_{O_2} – oxygen tension, p_g – boundary value of partial oxygen pressure, a_0, a_2 – constants.

The parameter a_0 in an algebraic model of the risk function $R\left(t, p_{O_2}\right)$ of *CNSyn* occurrence serves as a scaling factor. The boundary value of oxygen partial pressure p_g expressed in absolute atmospheres is the level below which $p_{O_2} < p_g$ the risk of *CNSyn* occurrence is not accumulated. Constant a_1 and a_2 being exponents enable modelling nonlinearity of the risk function $R\left(t, p_{O_2}\right)$. For $R\left(t, p_{O_2}\right)$ and $a_2 = 1$ the risk function is constant $R = const$. For $a_1 > 0$ the risk function $R\left(t, p_{O_2}\right)$ will grow linearly with the increase in oxygen partial pressure p_{O_2} or more than linearly if $a_1 > 1$. For $a_2 > 1$, the risk function $R\left(t, p_{O_2}\right)$ will increase more than linearly with the increase in exposure time t.

Hazard function[44] ξ, expressing the probability $P\left(t, p_{O_2}\right)$ of *CNSyn* symptom occurrence can be determined from Wiener relation (4.20) and equation (4.21) and written as: $\forall_{R\left(t, p_{O_2}\right) = a_0 \cdot a_2 \cdot (p_{O_2} - p_g)^{a_1} \cdot t^{a_2 - 1}} \xi \equiv F\left(t, p_{O_2}\right) = P\left(t, p_{O_2}\right)$ $= 1 - \exp\left[-\int_0^t R\left(t, p_{O_2}\right) dt\right.$. Since $\int b \cdot x^{a-1} dx \equiv \dfrac{b}{a} \cdot x^a$, the hazard ξ of *CNSyn* symptom occurrence for the single exposure procedure can be converted to the form:

$$\xi\left(t, p_{O_2}\right) \equiv F\left(t, p_{O_2}\right) = 1 - \exp\left[-a_0 \cdot \left(p_{O_2} - p_g\right)^{a_1} \cdot t^{a_2}\right] \tag{4.24}$$

Parameters $a_0..a_2$ and p_g for equation (4.24) were determined from experimental data (Harabin A.L., 1993) using the method of maximum credibility (Kłos R., 2007) – Figure 4.4, Figure 4.5 and Table 4.4.

Analyzing the data contained in Table 4.3, it can be noted that the model is not satisfactory, because for many parameters the uncertainty of their designation is greater than their set values. We found that the algebraic mathematical model simplified in relation to equation (4.23) of risk R of onset of *CNSyn* symptoms:

$$R(p_{O_2}) = a_0 \cdot (p_{O_2} - p_g)^{a_1} \tag{4.25}$$

gives a better approximation for the experimental data collected in Table 4.4 (Harabin A.L., Survanshi S.S., Homer L.D., 1995).

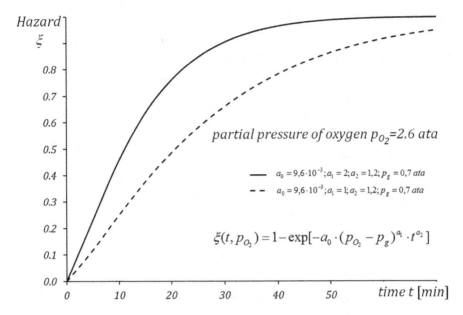

FIGURE 4.4 Hazard function x for dives of the single exposure type rest for the oxygen partial pressure $P_{O_2} = 0.26\,MPa$ and values of a_0, a_1 and p_{ig} estimated from the data for equation (4.24) – solid line, dotted line for $a_1 = 1$ (Yarbrough O.D., Welham W., Brinton E.S., Behnke A.R., 1947; Donald K., 1992).

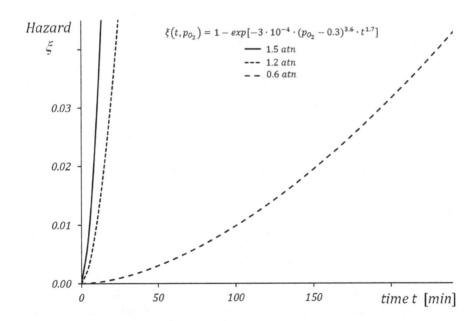

FIGURE 4.5 Hazard x resulting from model (4.24) of single exposure type dives performed in 1979–1986 (Harabin A.L., Survanshi S.S., Homer L.D., 1994).

TABLE 4.4

Parameters calculated with the method of maximum credibility for the risk ξ (4.24) of onset of *CNSyn* symptoms.

Type of data	Parameter			
	a_0	a_1	a_2	p_g
				[*ata*]
Immersed, under workload**	$(3\pm27)\cdot10^{-4}$	(3.6 ± 5.1)	(1.70 ± 0.33)	(0.3 ± 0.7)
Immersed, under workload*	$(4.7\pm4.7)\cdot10^{-3}$	(3.0 ± 1.8)	(1.27 ± 0.22)	(0.1 ± 0.4)
Immersed, without workload*	$(9.6\pm3.9)\cdot10^{-3}$	(2.1 ± 1.6)	(1.22 ± 0.21)	(0.7 ± 0.4)
Not immersed, without workload*	$(1.8\pm1.8)\cdot10^{-3}$	(3.0 ± 1.1)	(1.75 ± 0.14)	(1.4 ± 0.3)

The accuracy of the designation is given as a single standard deviation.

* tests were done before year 1972

** tests were done in years 1972–1994

In contrast to equation (4.23), the risk R of the onset of *CNSyn* symptoms (4.25) is not a function of time $R\left(p_{O_2}\right) \neq f(t)$.[45] Risk function $R(p_{O_2})$ in equation (4.25) is used to calculate the hazard ξ of the onset of *CNSyn* symptoms from all the observations contained in Table 4.4. When *CNSyn* causes abortion of exposure after the time T_1, the hazard $\xi_1\left(t, p_{O_2}\right) = F\left(t, p_{O_2}\right) = 1 - \exp\left[-\int_0^{T_1} R\left(p_{O_2}\right) dt\right]$ of *CNSyn* symptom onset is calculated according to the formula consistent with the Wiener relation (4.20):

$$\forall_{p_{O_2} \geq p_g} \ \xi_1\left(T_1, p_{O_2}\right) = 1 - exp\left[-a_0 \cdot \left(p_{O_2} - p_g\right)^{a_1} \cdot T_1\right] \tag{4.26a}$$

When the exposure ends after time T_0 without *CNSyn* symptoms, the hazard $\xi_0\left(t, p_{O_2}\right) = \exp\left[-\int_0^{T_0} R\left(p_{O_2}\right) dt\right]$ can be expressed as:

$$\forall_{p_{O_2} \geq p_g} \ \xi_0\left(T_0, p_{O_2}\right) = exp\left[-a_0 \cdot \left(p_{O_2} - p_g\right)^{a_1} \cdot T_0\right] \tag{4.26b}$$

The estimated values of $a_0..a_2$ and p_g refer to the partial pressure of oxygen p_{O_2}, which in equation (4.26) is expressed in $[p_g] = ata$, like the border oxygen partial pressure $[p_g] = ata$. Estimated value of a_0 is the scale coefficient while the border partial pressure p_g is the threshold beyond which $p_{O_2} > p_g$ and risk function R becomes larger than zero ($R > 0$). In the present algebraic mathematical model of hazard ξ of the onset of *CNSyn* symptoms, the border value of partial pressure p_g is always larger than or equal to 1: $p_g \geq 1$ *ata* – it is assumed that there are no signs of *CNSyn* symptoms below the partial pressure of O_2 at $p_{O_2} < 1\,ata$. For the estimated coefficient $a_1 = 0$ risk function R is constant, $R = const$, independent of the value of oxygen partial pressure p_{O_2}. When the estimated coefficient $a_1 = 1$, the risk function

R increases linearly together with the increase in oxygen partial pressure P_{O_2}. For the $a_1 > 1$, this increase is faster than linear.

For the fixed value of the border oxygen partial pressure $p_g \equiv 1\ ata^{46}$ (Table 4.5), parameter values from algebraic mathematical model of hazard ξ of $CNSyn$

TABLE 4.5
Experimentally tested oxygen dive profiles, which were used to determine the hazard ξ of the onset of *CNSyn* symptoms (Harabin A.L., Survanshi S.S., Homer L.D., 1995).

Profile no.	Exposure profiles with places marked where symptoms of *CNSyn* occurred						n /N
	Time [*min*]/Depth [*fsw*]						
Exposures with the change of depth							
1	60/25	15/40[†]	60/25				1/14
2	120/25[‡]	15/40[††]					3/15
3	120/20[†]	25/35[†]	95/20				2/20
4	240/20[††]	25/35[†]					3/19
5	120/20	25/35[†]	95/20	25/35[†]			2/16
6	120/20	15/40[†]	105/20				1/39
7	240/20	15/40[‡]					1/16
8	120/20	10/50[†]	110/20				1/18
9	15/40	30/20	15/40[††‡]				3/24
10	15/40	45/25	15/40[†]				1/4
11	15/40[†]	60/20	15/40[††‡]				4/47
12	15/40	90/20[†††‡]	15/40	90/20	15/40	15/20	4/11
13	15/40[††]	225/20[††]					4/64
Total							30/307
Exposure to single depth							
14	154/25[*]						0/12
15	240/25						0/22
16	129/30[*††††]						4/18
17	90/30[‡]						1/40
18	25/35						0/47
19	30/35[†††††]						5/40
20	15/40						0/70
21	20/40[‡‡]						2/17
22	5/50						0/57
23	10/50						0/58
Total							12/381
Cumulative total							42/688

Cases of absence of *CNSyn* symptoms described in publications by various researchers were collected, also occurrence of symptoms [†], and occurrence of seizures [‡], time and depth of dive of exposure end was given

Exposure numbers 1–13 are shallow water profiles with 1–3 trips to depth of {35; 40; 50} *fsw* with the time of stay in a range of 10–25 *min* while 14–23 are exposure to single depth.

N – total number of exposures
n – number of cases of occurrence
[*] – approximate time of the exposure

occurrence, estimated with the maximum credibility method, were $a_0 = (1.33 \pm 0.22) \cdot 10^{-3}$ and $a_1 = (3.39 \pm 0.5)$.

Most of the research results used in the process of reaching conclusions were obtained in the course of investigations carried out in warm water 22°C.[47] The diver was in a fetal position[48] performing a cyclic task[49] with the use of cycle ergometer in shallow water with an average load – oxygen consumption averaged to $\dot{v}_0 \approx 1.3 \, dm^3 \cdot min^{-1}$. Close attention was paid to the content of oxygen and carbon dioxide, trying to keep them at the appropriate levels: $C_{O_2} > 95\%$ and $C_{CO_2} < 1\%$.[50]

Exposures during which the diver dissimulated[51] and on completing them claimed that he had *CNSyn* symptoms were excluded from the analysis. The symptoms regarded as finishing the exposures are nausea, numbness, dizziness, cramps, impaired hearing and vision, loss of consciousness and convulsions.[52]

The results of matching the algebraic models of risk $R(p_{O_2})$ as expressed by equation (4.25) and the danger $\xi\left(T, p_{O_2}\right)$ of the onset of *CNSyn* symptoms, expressed by equation (4.26), are shown in Table 4.6.

Hazard $\xi\left(T, p_{O_2}\right)$ of the onset of symptoms of *CNSyn* for single exposure type dives[53] was calculated directly from the model (4.26) (Figure 4.6).

For the multilevel dives, danger was calculated separately for each level and then those values were added. Examples of calculation for profile 13 from Table 4.6 are shown in Figure 4.7.

Number n of predicted events with occurrence of *CNSyn* was calculated for each profile from Table 4.5 by multiplying the values obtained for hazard $\xi\left(T, p_{O_2}\right)$ by the number of experimental dives N (Table 4.6).

The presented method of modelling and predicting the hazard ξ of onset of *CNSyn* symptoms raises no objections as regards the profiles of single type exposure. However, for multilevel profiles, it assumes that there is no difference between the duration sequences of stay at the individual depths. For example, there is no difference between exposures 6 and 13 from Table 4.5.

According to the theory of the biochemical mechanisms of *CNSyn* symptoms occurrence, there should be a difference between the profile in which the trip to the greater depth is at the beginning of the transit and that in the middle or at the end of exposure. It is possible to conceive of a situation in which the statistical average production of harmful metabolites during multilevel oxygen exposure poses a relatively low risk, as in the case under study. Increased production of the harmful radicals during a trip is stopped, gradually purged or relaxed during a stay at a smaller depth. This effect, however, can be small when oxygen exposure continues, in comparison to the effects produced during rest under norm-baric conditions. Hence, for exposures that differ in the moments the trip starts, there should occur different concentrations of harmful radicals after a dive, giving differences in the initial state for the repetitive dive. If, however, the time until the next dive is long enough to complete full deactivation of harmful metabolites, such exposures are indistinguishable when this method of inference is used.

TABLE 4.6

Comparison of the danger ξ of onset of *CNSyn* symptoms defined using the algebraic semi-empirical mathematical models (4.26) with the binominal distribution (Harabin A.L., Survanshi S.S., Homer L.D., 1995).

Profile no. from Table 4.5	No. of observations	No. of CNSyn cases	Binominal distribution			Algebraic model (4.26a)	
			Confidence interval[+]		Predicted number of CNSyn	Hazard of CNSyn	Predicted number of CNSyn
	N	n	p_1	p_r	n_b	ξ	n
Exposures with the change of depth							
1	14	1	0.0018	0.3387	0.0–4.7	0.0989	1.4*
2	15	3	0.0433	0.4809	0.6–7.2	0.0979	1.5*
3	20	2	0.0123	0.3770	0.2–7.5	0.0915	1.8*
4	19	3	0.0338	0.3958	0.6–7.5	0.0966	1.8*
5	16	2	0.0155	0.3835	0.2–6.1	0.1312	2.1*
6	39	1	0.0006	0.1348	0.0–5.3	0.0916	3.6*
7	16	1	0.0016	0.3023	0.0–4.8	0.0943	1.5*
8	18	1	0.0014	0.2729	0.0–4.9	0.1082	1.9*
9	24	3	0.0266	0.3236	0.6–7.8	0.0485	1.2*
10	4	1	0.0063	0.8059	0.0–3.2	0.0982	0.4*
11	47	4	0.0237	0.2038	1.1–9.6	0.0897	4.2*
12	11	4	0.1093	0.6921	1.2–7.6	0.1597	1.8*
13	64	4	0.0173	0.1524	1.1–9.8	0.0909	5.8*
Exposure to single depth							
14	12	0	0.0000	0.3187	0.0–3.8	0.0768	0.9‡
15	22	0	0.0000	0.1889	0.0–4.2	0.1171	2.6‡
16	18	4	0.0641	0.4764	1.2–8.6	0.1168	2.1‡
17	40	1	0.0006	0.1316	0.0–5.3	0.0830	3.3‡
18	47	0	0.0000	0.0933	0.0–4.4	0.0398	1.9‡
19	40	5	0.0419	0.2680	1.7–10.7	0.0475	1.9‡
20	70	0	0.0000	0.0637	0.0–4.5	0.0376	2.6‡
21	17	2	0.0146	0.3644	0.2–6.2	0.0498	0.8‡
22	57	0	0.0000	0.0776	0.0–4.4	0.0268	1.5‡
23	58	0	0.0000	0.0763	0.0–4.4	0.0529	3.1‡

[+]according to (Kłos R., 2007):

$$\forall_{0<n<N}\begin{cases} P\left(\rho \geq p_r\right) = \sum_{x=0}^{n}\binom{N}{x}\cdot p_r^x \cdot (1-p_r)^{N-x} = \dfrac{\alpha_0}{2} \\ P\left(\rho \leq p_l\right) = \sum_{x=n}^{N}\binom{N}{x}\cdot p_l^x \cdot (1-p_l)^{N-x} = \dfrac{\alpha_0}{2} \end{cases};$$

TABLE 4.6 *(Continued)*
Comparison of the danger ξ of onset of *CNSyn* symptoms defined using the algebraic semi-empirical mathematical models (4.26) with the binominal distribution (Harabin A.L., Survanshi S.S., Homer L.D., 1995).

$$\forall_{n=0}\begin{cases} p_l = 0 \\ p_r = 1 - 10^{\lg\frac{\alpha_0}{N}} \end{cases}; \quad \forall_{n=N}\begin{cases} p_l = 10^{\lg\frac{\alpha_0}{N}} \\ p_r = 1 \end{cases}$$

$^{\ddagger} n = N \cdot \{1 - \exp[-1{,}33 \cdot 10^{-3} \cdot (p_{O_2} - 1)^{3.39} \cdot t]\}$

N – general number of exposures from Table 4.5
t –exposure time from Table 4.5
n – number of *CNSyn* cases
* adding the values calculated according to formula from the note ‡

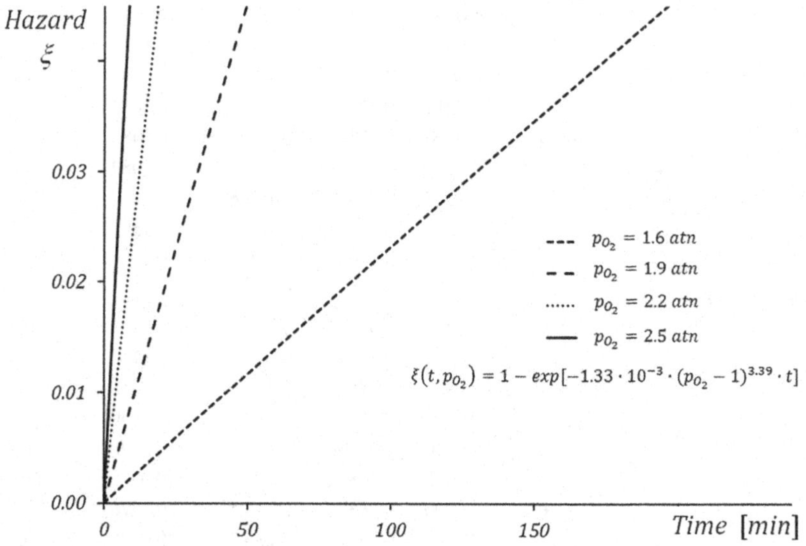

FIGURE 4.6 Hazard ξ for single type exposure dive defined according to model (4.26a).

4.15 RISK COMPARISON

According to model (4.26), there is no difference in predicting hazard ξ of *CNSyn* onset of symptoms. This is confirmed by the small difference in values of profiles 6 and 13. If there were a difference between profiles 6 and 13, it could be manifested by a range of prediction of *CNSyn* cases calculated from a binominal distribution (Table 4.6).

When comparing the risk values, a set of hypotheses can be put forward:

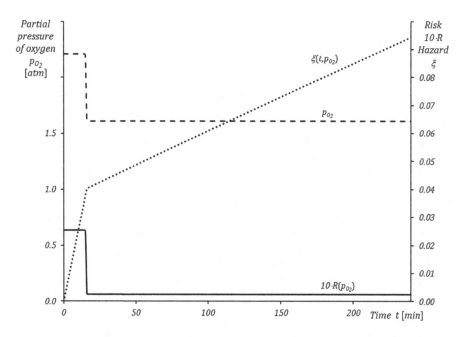

FIGURE 4.7   The dependence of the cumulative hazard ξ and risk R on time t for multilevel dive no. 13 from Tables 4.5 and 4.6 (Harabin A.L., Survanshi S.S., Homer L.D., 1995).

$$\begin{cases} H_0 : \xi_6 = \xi_{13} \\ H_0 : \xi_6 \neq \xi_{13} \end{cases} \qquad (4.27)$$

From Table 4.5 we learn that for profile 6 with the number of performed tests $N_6 = 39$, there was $n_6 = 1$ – one case of *CNSyn* symptoms; with $N_{13} = 64$ performed tests there were $n_{13} = 4$ cases of *CNSyn* symptoms for profile 13. To verify whether different moments to begin trips have an influence on the risk of onset of *CNSyn* symptoms the probability of the results of experiments can be calculated, assuming that the null hypothesis H_0 is true.

The results of the experiments can be treated as samples of two populations.[54] Because the average for the two populations is not known, they should be estimated from the sample. Assuming respectively that the value of random variable is $X \leftarrow 1$ if symptoms of *CNSyn* occur and $X \leftarrow 0$ as the case of absence of *CNSyn* symptoms, according to the definition the estimation of expected value $EX_i \equiv \Sigma_i X_i \cdot P_i$ from the sample i for binominal distribution, the estimator of hazard ξ_i can be written as: $\hat{\xi}_i = \dfrac{1 \cdot n_i + 0 \cdot (N_i - n_i)}{N_i} = \dfrac{n_i}{N_i} \equiv \bar{x}_i$, where $\bar{x}_i$ is the average value from the sample (Kłos R., 2007). From here estimators of expected values for each profile can be calculated: $\hat{\xi}_6 \equiv \bar{x}_6 = \dfrac{n_6}{N_6} = \dfrac{1}{39} \cong 0.0256$ and $\hat{\xi}_{13} \equiv \bar{x}_{13} = \dfrac{n_{13}}{N_{13}} = \dfrac{4}{64} \cong 0.0625$.

For the binominal distribution the variance $D^2 X \equiv N \cdot x \cdot (1 - \xi)$ can be estimated from the sample: $D^2 X \triangleq s_i^2 = N \cdot \hat{\xi_i} \cdot \left(1 - \hat{\xi_i}\right)$ (Kłos R., 2007). Hence, the variance of profile 6 will be: $s_6^2 \cong 39 \cdot 0.0256 \cdot (1 - 0.0256) \cong 0.9728$ and for profile 13: $s_{13}^2 \cong 64 \cdot 0.0625 \cdot (1 - 0.0625) \cong 3.755$. The standard deviation of the average value can be calculated according to the equation: $s_{\bar{x}_i} = \sqrt{\hat{s}_i^2 / N}$. For each trial, they amount to: $s_{\bar{x}_6} \cong \sqrt{\frac{0.9728}{39}} \cong 0.1579$ and $s_{\bar{x}_{13}} \cong \sqrt{\frac{3.75}{64}} \cong 0.2421$.

Formally, the sample size is sufficient to approximate the binominal distribution with the normal distribution. However, since the size of one of the groups only slightly exceeds 30, the *t-test* was applied.

The calculations were made for the average distribution of the estimated hazard $\hat{\xi_6}$ expressed by binominal distribution. It has been approximated by $t - Student$ distribution for profile 6 of Table 4.5, for which sample $N_{13} = 64$ decompression profiles were taken at random. This resulted in adopting $v = N_{13} - 1 = 63$ degrees of freedom for the adopted $t - Student$ distribution. The interesting question was, "What is the probability for the estimated average danger $\hat{\xi}_{13}$ to occur for profile 13 having the same value as that observed for profile 6 or higher $\hat{\xi}_{13} \geq \hat{\xi}_6$."

Using the formula $t = (\bar{x} - \mu) : \sqrt{s^2 / N - 1}$ referred to as the statistics, where: $\bar{x}$ represents average from sample, μ is the true value of binominal distribution describing the sample, s^2 represents sample variance, N the size of the sample, the probability a_0 of the statistical distribution of sample 6 can be calculated, wherein the estimated hazard x is the same or greater than the estimated hazard for distribution 13. For this purpose, the distribution for sample 6 is treated as a reference. Hence, for the statistics t value of $\mu \equiv \bar{x}_6$, $s = s_{\bar{x}_6}$, $N = 40$, while the test value will be $\bar{x} = \bar{x}_{13}$:

$t = \frac{\bar{x}_{13} - \bar{x}_6}{s_{\bar{x}_6}} \cong \frac{0.0625 - 0.0256}{0.1579} \cong 0.2337$. For the $t - Student$ distribution value of statistic $t \cong$ 0.2337 corresponds to the value of probability $a_0 (N_{13} - 1 = 63) \cong 0.40800$.

From the performed calculations we can see that of all 64 samples of the profile 6 population, which could be selected at random, approximately 41% have the same or higher value of the estimated risk as in the case of the profile $\hat{\xi}_{13}$. Based on the studies carried out and estimates made, it cannot be concluded that the estimated risks ξ of *CNSyn* symptoms for both populations and the exposure distribution applied differ significantly.[55] The result justifies the way accepted for estimating risk, i.e. without taking into account the sequence of trips made.

4.16   SUMMARY

The methods of survival analysis were introduced to the problems associated with diving by Weathersby and Thalmann (Gerth W.A., 2002). The hazard prediction model ξ of *CNSyn* symptoms, proposed by the US Navy, which derives from this theory, seems to be sufficiently precise. Its strength is that several researchers have confirmed the same border value of oxygen partial pressure $p_g \equiv 1\ ata$, beyond which the danger of *CNSyn* is expected. This limits the safe exposure time when breathing a medium containing oxygen under partial pressure above the magnitude defined by this border value $P_{O_2} > P_g$.

In the proposed algebraic mathematical model, the dose of *CNSyn* toxicity accumulates only during the dive. The dose of *CNSyn* toxicity is independent of the sequence of the dive phases.[56]

However, it seems that despite identical durations of individual phases of the dive, there should be a difference between the profile which, for example, starts exposure with a trip to a greater depth and the profile in which the trip is undertaken at the end of dive. The results of the tests allowed for statistically comparing two profiles for differences in the risk run at different times of the trip. The statistical inference process applied to them has not given any grounds for rejecting the null hypothesis, which refers to an absence of statistically significant differences in risk carried when changing the sequence of trips.

NOTES

1 For example, the analysis of time from the start of treatment until disease recurrence, death, etc.
2 For example, time of unemployment, primiparity (a medical term used to refer to a condition or state in which a woman is bearing a child for the first time and/or has given birth to an offspring at one time).
3 For example, time of damage of the piece of equipment, uptime, etc.
4 Depending on the scheme in question, it can be called the time of trouble-free operation, duration, expectations or answers.
5 For example, in estimating survival of a heart attack, the starting point may be related to the time of onset of symptoms, admission to hospital, start of a particular treatment, etc.
6 Such as day, hour, minute, etc.
7 For example, in reliability engineering it can be expressed in a number of cycles performed by a machine until failure occurs.
8 For example, for the two alternative methods of treatment.
9 Data are called cutoff when their value cannot be accurately estimated – this will be discussed later.
10 For example, binominal, normal, F, χ^2, t, etc.
11 In typical statistical inference, one would have to wait until the patient dies, but this may take many years or decades, i.e. sometimes it is unrealistic to wait for the established endpoint of the study to occur.
12 For example, for dead patients the survival time T can be defined.
13 That is, survival time T is greater than the end point of test t – it is known that survival time T is within the range (t; ∞).
14 A patient may be withdrawn from the program of research, contact with him may be lost at the end of the test cycle, there is no economic or practical justification of patient monitoring until a planned final event occurs, etc.
15 Very frequent irregularities in practice.
16 Sufficient conditions for data not to carry significant information about the nature of the phenomenon is independence of survival time T from final time t.
17 As a result of illness or loss of communication with the cured patient, or as a result of his/her failure to turn up for appointments.

18   Similarly, taking into consideration data "incorrectly" cut off may lead to inadequate conclusions.

19   Some technical systems (e.g. pressure vessels, elevators, cranes) require a periodic certification process, after the period of permitted operation, the device is usually suitable for safe use, but despite its technical efficiency, its exploitation is prohibited by law.

20   Generalized distribution Γ – Table 4.1.

21   For example, the onset of *CNSyn* symptoms.

22   For example, at N time periods.

23   As it will be shown later, the start of thinking about the inverse event simplifies the introduction of cumulative distribution of risks.

24   For example, risk of the onset of *CNSyn* symptoms.

25   There is always a possibility of even a small risk, thus postulating risk values at zero $R \equiv 0$ is in contradiction with observed reality.

26   $\forall_{f>0} \int_0^\infty L(t,f) \cdot dt = f \cdot \int_0^\infty e^{-f \cdot t} \cdot dt = f \cdot \dfrac{1}{-f}\Big|_0^\infty = 0 - \dfrac{f}{-f} = 1.$

27   When the lifetime was so far t.

28   Remaining lifetime Δt does not depend on the past and has the same exponential distribution, as up to now survival time is t.

29   As a rule, people die and the machines start to deteriorate after reaching a certain age.

30   For example, during a single diving cycle.

31   The probability of failure-free operation, of survival, of any other defined event, etc.

32   For example, $S(t)$ is the probability that a given person will live up to time t.

33   In engineering an equivalent of survival function $S(t)$ is used to determine the reliability and is called safety and *reliability function*.

34   The value of hazard function $h(t)$ should be considered as a potential for occurrence of the expected event (mostly unsuccessful) showing that the analysis of the problem situation characterized by the survival function $S(t)$ has been completed; when the function $S(t)$ decreases, the $h(t)$ grows. The function $h(t)$ can be graphically compared to the operation of car speedometer. On the basis of its constant it can be concluded what distance will be covered after the selected time has elapsed; on the basis of the fixed value of the function $h(t)$, the number of expected events within the selected time can be concluded.

35   In engineering safety, the hazard function $h(t)$ is defined as *intensity of safety failure* and is often written as $\lambda_B(t)$.

36   Depending on the adopted units of time.

37   One important reason for using the hazard function $h(t)$ is that the conditional distribution of the expected survival time beyond the moment t_0 can be calculated directly from it for $h(t > t_0)$.

38   For example, a measure of reliability is the probability of meeting the system's requirements per unit of time, while a measure of hazard is the probability of an adverse situation occurring for the space-time surrounding the system.

39   Depending on the adopted explication.

40   Using the formula: .

41   Search for the dominant using the method of maximum credibility.

42   Constant c has the inverse of a unit of time; in this case it may be $[c] = h^{-1}$.

43   Dependence equivalent to the risk function , which defines the probability of *CNSyn* occurrence for the procedure of single exposure type, in conjunction with exposure to oxygen partial pressure , to which a diver is exposed for time t.

44   Which is a cumulative distribution function F.

45   Risk function R describing the intensity of the *CNSyn* occurrence at a selected time in (4.25) is constant value $R = const$ independent of the time for oxygen partial pressure ; the *CNSyn* risk dependency on time occurs when hazard is being calculated.

46 Estimated value of the border partial pressure for data from Table 4.5 was $p_g = (1.3 \pm 0.4)$, but the value $p_g \equiv 1$ *ata* was adopted as more physiological.

47 The tests carried out in water at 12°C and 4°C were not taken into consideration because they pose an additional threat identified earlier by Donald (Donald K.W., 1992).

48 A cycle ergometer was submerged in a shallow container set in a hyperbaric chamber so that the diver leaning while sitting on a cycle ergometer at any time could raise his head and after removing the mouthpiece could start to breathe the air in the chamber (Harabin A.L., Survanshi S.S., Homer L.D., 1995).

49 6 *min* work and 4 *min* rest (Harabin A.L., Survanshi S.S., Homer L.D., 1995).

50 It was not specified whether this tension was related to the atmospheric pressure.

51 This is understood here as presence manifested in an attempt to meet the norms or hide the actual situation and feelings, impulses, behaviors, etc.

52 The frequency distribution of particular symptoms is shown in Table 3.2.

53 Exposures at one value of the partial pressure of oxygen throughout the experimental diving – profiles 14–23 from Tables 4.5 and 4.6.

54 Infinitely many such samples can be taken from them, and the distribution of averages from such samples will tend, following the central limit theorem, towards a normal distribution with the population average estimated as the average of the sample and a standard deviation estimated as the standard deviation of the sample divided by the square root of the sample size (Kłos R., 2007).

55 Running a formal *t-test* for unrelated samples has the same effect, which says that there are no grounds to reject the null hypothesis H_0 (4.27) at the significance level of $a_0 = 5\%$.

56 Hazard of *CNSyn* for profiles with a trip to the greater depth does not depend on the time of the trip.

REFERENCES

Brubakk AO & Neuman TS. 2003. *Bennett and Elliott's Physiology and Medicine of Diving.* London: Saunders. ISBN 0-7020-2571-2.

Donald KW. 1992. *Oxygen and the Diver.* Harley Swan: The SPA Ltd. ISBN 1-85421-176-5.

Evans M, Hastings N & Peacock B. 1993. *Statistical Distributions.* New York: John Willey & Sons, Inc. ISBN 0-471-55951-2.

Gerth WA. 2002. Overview of survival functions and methodoology. In *Survival Analysis and Maximum Likelihood Techniques as Applied to Physiological Modeling,* ed. PK Weathersby & WA Gerth, 1–28. Kensington: Undersea and Hyperbaric Medical Society Inc.

Harabin AL. 1993. *Human Central Nervous System Oxygen Toxicity Data from 1945 to 1986.* Bethesda: Naval Medical Research Institute. NMRI 93-03; AD-A268-225.

Harabin AL, Survanshi SS & Homer LD. 1994. *A Model for Predicting Central Nervous System Toxicity from Hyperbaric Oxygen Exposure in Man: Effects of Immersion, Exercise, and Old and New Data.* Bethesda: Naval Medical Research Institute. NMRI 94-0003; AD-A278 348.

Harabin AL, Survanshi SS & Homer LD. 1995. A model for predicting central nervous system toxicity from hyperbaric oxygen exposure in humans. *Toxicol. Appl. Pharmacol.,* 132, 19–26.

Hills BA. 1997. *Decompression Sickness – The Biophysical Basis of Prevention and Treatment: Vol. 1.* Chichester: John Willey&Sons. ISBN 0-471-99457-X.

Kłos R. 2007. *Zastosowanie metod statystycznych w technice nurkowej - Skrypt.* Gdynia: Polskie Towarzystwo Medycyny i Techniki Hiperbarycznej. ISBN 978-83-924989-26.

Wienke BR. 2003. *Basic Decompression Theory and Application*. Flagstaff: Best Publishing Co. ISBN 1-930536-14-3.

Yarbrough OD, Welham W, Brinton ES & Behnke AR. 1947. *Symptoms of Oxygen Poisoning and Limits of Tolerance at Rest and at Work*. Washington: Naval Experimental Diving Unit. Research Report 01-47.

5 Oxygen Exposures

Guidelines for handling oxygen exposures were based on the best knowledge obtained during our own research and on analysis of the literature. To develop them, the US Navy technology of diving, based on the studies carried out in the 1980s (Butler F.K., Thalmann E.D., 1984; Butler F.K., Thalmann E.D., 1986a; Butler F.K., Thalmann E.D., 1986b), was taken into consideration. Our own studies showed inconsistencies in some of the US Navy procedures. As a result, some modifications of these procedures were proposed.

5.1 TACTICAL CONSIDERATIONS

Oxygen combat diving is used to perform covert transfer of special operations forces groups/sections. As a standard, they are not very deep, not exceeding 6 mH_2O. Diving depth is limited not only by the toxic effects of oxygen, but also by tactical missions. Technical passive[1] means used to detect combat divers are usually placed at the bottom. Their performance drastically decreases with distance. Keeping close to the surface minimizes the risk of detected by these technical measures.[2]

To provide protection against divers, specialized active and passive sonar systems are used. They are designed specially to detect divers.[3] Active sonars used to detect divers have less detection effectiveness close to the surface,[4] especially when there is a bigger target on the course of sonar sweep, such as a surface vessel, floating garbage, fauna, a layer of suspended air formed in the course of waving and so on.

The diver is an object with a relatively low target reflectivity[5] and small size. Therefore, these devices are characterized by limited effective range, which depending on the hydrological conditions is [300, 800]m (Pozański P., 2011). It has been established that 800 m was adopted here as the smallest operating radius for approaching the shore without prior reconnaissance because of the possible existence of shore protective/reconnaissance devices.[6] Taking into consideration the aforementioned analysis, it was assumed that the excursion phase[7] occurring after an oxygen dive to a depth of 6 mH_2O can be expected, at the earliest, after 45 min of transfer.

The drawback of staying close to the surface in clear water is that the diver can be relatively easily detected by observers on the surface. However, because of water reflections, such observers need to be relatively high and should be looking at such an angle that these reflections do not blind them. Use of the polarizers improves surveillance conditions because reflected light is significantly polarized. Promising results are expected from using the green laser to detect shallow-water objects.

In muddy water, in the absence of breathing medium emissions, a slowly swimming diver is a target difficult to detect providing that his fins do not produce water swirling on the surface. Therefore, combat diving operations must take into consideration training in close-to-surface swimming at a relatively low speed. The analysis presented here was the basis for planning scenarios for experimental dives.

DOI: 10.1201/9781003309505-6

5.2 VARIANTS OF OXYGEN DIVES

The basic definitions used herein are collected in Table 5.1.

The basis for a new generation of diving methods with oxygen as a breathing medium was provided by the studies carried out in the 1980s (Butler F.K., Thalmann E.D., 1984; Butler F.K., Thalmann E.D., 1986a; Butler F.K., Thalmann E.D., 1986b). Two categories of diving operations were then introduced (US Navy Diving Manual, 2008; US Navy Diving Manual, 2016):

- transit at a shallow depth; the diver is allowed to make one excursion to greater depths, hereinafter referred to as the *transit with excursion limits* procedure (Table 5.2).
- single-depth oxygen exposure limits, hereinafter referred to as *single-depth limits* procedure (Table 5.3).

The transit with excursion limits is normally the preferred mode of operation because maintaining a depth of 6 mH_2O or shallower minimizes the possibility of *CNSyn* symptoms occurrence.

TABLE 5.1
The terms used in diving procedures using oxygen and Nx.

Specification	Definition
Transit with excursion limits	Diving to a maximum dive depth of 6 mH_2O or shallower allows the diver to make a brief excursion to depths as great as 15 mH_2O
Transit	The portion of the dive spent at a depth of 6 mH_2O or shallower
Excursion	The portion of the dive spent at a depth greater than 6 mH_2O, but shallower or equal to 15 mH_2O
Excursion time	The time between the diver's initial descent below 6 mH_2O and returning to 6 mH_2O
Single-depth limits procedure	In this procedure the diver is allowed to make a dive to a depth of 15 mH_2O
Repetitive oxygen diving	Diving with a closed circuit apparatus with oxygen as a breathing medium, when an interval between single dives is less than 2 h
Off-oxygen interval	The time from when the diver discontinues breathing oxygen on one dive until he/she begins breathing oxygen again from the closed circuit apparatus, having oxygen as a breathing medium, on the next dive

---

TABLE 5.2
Excursion limits (US Navy Diving Manual, 2008; US Navy Diving Manual, 2016).

Depth	Maximum time
[mH_2O]	[*min*]
6–12	15
12–15	5

TABLE 5.3
Single-depth oxygen exposure limits (US Navy Diving Manual, 2008; US Navy Diving Manual, 2016).

Depth	Maximum oxygen time
[mH₂O]	[min]
6	240
9	80
10.5	25
12	15
15	10

Comparing the data from Table 5.2 and Table 5.3, it can be seen that the theories underlying each of them were different. *The* single-depth limits (Table 5.3) allow maximum exposure at a depth of 12 mH_2O to be 15 *min*, and the dive should be aborted. However, in the case of the transit with excursion limit, after reaching a dive depth of 12 mH_2O and the excursion time of 15 *min*, the diver is allowed to continue the dive to the exposure time of 240 *min*.

The proposed modifications have been based on the studies conducted by the US Navy[8] and partly in the Polish Naval Academy (AMW).[9]

The oxygen exposure limits described here are applicable to closed circuit oxygen apparatus *CCR – SCUBA* dives with oxygen as a breathing medium. The results confirmed the usefulness and safety of the procedures observed by a group of specially trained combat divers. In practice, however, individual cases of increased sensitivity to the toxic effects of oxygen on the central nervous system or respiratory system may occur (Donald K.W., 1992). For this reason during such dives maximum safety measures, appropriate for the specific circumstances, should be taken. In the Polish Navy, it is assumed that dives with oxygen and artificial breathing mediums are allowed only for trained divers who have successfully passed the oxygen tolerance test (see section 3.4 and Appendix 2).

5.3   MODIFIED TRANSIT TYPE PROCEDURE

The transit with excursion limits is the most common procedure during combat special forces operations. If the task requires diving to a depth greater than 6 mH_2O and exposure longer than the permissible excursion duration, than the appropriate single-depth limit should be used. The diver, who is at a depth of 6 mH_2O or less, can make a deeper excursion, but preparation for the excursion should be calculated according to the following rules:

- maximum total time of dive[10] may not exceed 240 *min*,[11] and when the allowed deeper excursion is made, it is reduced to 120 *min* plus the time of the excursion
- only one excursion is allowed to a depth within the range (6; 15) mH_2O
- the time permitted for an excursion is limited due to the maximum depth reached during the excursion (Table 5.4)

TABLE 5.4
The maximum allowable exposure oxygen times and sustainable risks $\xi < 5.5\%$, according to equation (4.26a).

Depth range	Maximum allowable excursion time	Remarks
[mH_2O]	[min]	
Oxygen		Maximum exposure time up to 240 min
6–12	10	After an oxygen excursion maximum exposure time
12–15	5	decreases to 120 min
Nitrox 0.43 O_2/N_2		The last 5 min is treated as an emergency time needed to
6–24	30	equalize potential delays

- a single excursion can start at any time while the diver is staying at the depth of transit, but its execution limits the time of further transit to 120 min and no other excursion can be undertaken
- the diver must have returned from the excursion to a depth of 6 mH_2O or shallower by the end of the prescribed excursion limit

If an inadvertent excursion should occur during a dive, one of the following procedures should be applied:

a. if the depth and/or time of the excursion exceeds the limits or if an excursion has been taken previously, the dive must be aborted and divers must return to the surface
b. if on completion of the excursion total transit time has exceeded 120 min, the dive should be aborted
c. if an inadvertent excursion is the first one being made and has not exceeded limits contained in Table 5.4, the dive can be continued up to a maximum transit time of 120 min, and no extra excursions to a depth greater than 6 mH_2O are allowed
d. if an inadvertent excursion exceeds the transit with excursion limits, but the excursion is within the allowed single-depth limits, the dive can be continued according to this procedure in Table 5.5
e. if the diver makes the uncontrolled drop and is not sure how long he has stayed at a depth below 6 mH_2O, he should abort the dive

5.4   MODIFIED PROCEDURE FOR SINGLE-DEPTH OXYGEN EXPOSURE LIMITS

The permissible single-depth oxygen exposure limits are collected in Table 5.5. The diver does not have to spend the entire *dive at a single depth*. The maximum permitted dive time limit is a function of the maximum depth attained during the dive. In

TABLE 5.5
Single-depth oxygen exposure limits with balanced
$\xi \cong 5.5\%$ **hazard according to (4.26).**

Maximum diving depth	Maximum oxygen time
$[mH_2O]$	$[min]$
6	240
7	142
8	90
9	60
10	42
11	30
12	22
13	17
14	13
15	10

contrast to the transit with excursion limits, in the single-depth exposure type procedure extra excursions are not allowed.

5.5   EXCEPTIONAL EXPOSURE DIVES

During combat missions, unique diving situations may arise. They include repetitive dives. If a planned mission with oxygen involves repetitive diving, the impact of the previous dive on the permitted time and depth of the mission depends on the interval between the previous and the planned dives. If the oxygen diving is repetitive, then the oxygen exposure limit is determined according to the rules given in Table 5.6.

As regards the transit with excursion procedure, the daily limits listed in Tables 5.4 and 5.5 should not be exceeded. If the time interval between the previous dive and the planned successive dive is greater than 2 h, according to the single-depth exposure limits, the planned dive is regarded as an initial dive, provided that the total time limits, as set in Table 5.5, are not exceeded.

An oxygen dive undertaken after an off-oxygen interval of more than 2 h is considered the same as an initial oxygen exposure. If a negative number is obtained when adjusting the single-depth exposure limits as shown in Table 5.5, a 2 h off-oxygen interval must be taken before the next. Another unique situation encountered during combat missions is oxygen diving after other types of exposure. There is no simple procedure that can be applied to calculating permissible oxygen exposure limits for dives conducted after previous dives on air or mixed gas. If a diver uses a closed circuit oxygen apparatus $CCR - SCUBA$ for one portion of the dive and another apparatus with a breathing medium other than air for the second portion, only the portion of the dive made on oxygen is counted as oxygen exposure time.[12]

TABLE 5.6
Methods for calculating the permissible oxygen exposure limits for successive dives.

	Method of calculating maximum successive dive time	Excursion to a greater depth
Transit with excursion limits	Subtract time on previous exposures from 240 *min* and the time left is the permitted time for successive dive	Allowed if such an excursion was not made on previous dive and the transit time did not exceed 120 *min*
If none of the described situations cannot be applied, the diver has to wait at least 2 *h* before his next dive		
Single-depth oxygen exposure limits	1. Determine maximum oxygen dive time for deepest depth; diving depth is selected after analyzing previous and planned dives 2. Subtract oxygen time on previous dives from maximum successive dive time defined in pt. 1 above	No excursion allowed when using single-depth limits to calculate remaining oxygen time
If none of the two situations described in the table cannot be applied to the single-depth limits, the diver has to wait at least 2 *h* before his next dive		

Different types of diving apparatuses can be used during special forces operations using the technologies specially developed and tested for each tactical scenario.[13] If on previous dives[14] the diver was exposed to oxygen partial pressure Po_2 equal to or greater than $p_{O_2} \geq 0.1\,MPa$, the exposure must be considered when a successive oxygen dive is planned.

The interval between dives is calculated from the time of aborting breathing the previous mixture until the moment when the diver begins to breathe oxygen.

All of these procedures of diving with the use of oxygen as a breathing medium can be used[15] during diving operations conducted in reservoirs lying at high altitudes. Air transport immediately after diving is prohibited if diving with the use of oxygen was a part of the diving operation, during which at various depths other breathing media were also used.[16] In such a case, the possibility of air transport should be considered individually for each operation. However, it should be considered that the immediate implementation of air transport of divers after diving with closed circuit and oxygen as a breathing medium can lead to *DCS* (Donald K.W., 1955; Donald K.W., 1992).

5.6 SUMMARY

Most of the data used to develop the system described were obtained in the course of experimental dives in a water temperature range of $t \in [65, 70]°F$[17] (Harabin A.L., Survanshi S.S., Homer L.D., 1994). In most missions carried out by Polish combat divers the water temperature is lower,[18] and the effects of both supercooling and overheating on reduced tolerance to hyperoxia is significant (Donald K.W., 1992; Harabin A.L., Survanshi S.S., Homer L.D., 1994). Therefore, the research focused on this aspect of the most likely future combat missions.

During exposure, attention should be paid to the fact that the increased content of carbon dioxide CO_2 in the inhaled breathing medium increases the risk of *CNSyn*. Carbon dioxide CO_2 content in tissues does not increase only because of its recirculation resulting from incomplete absorption in the canister or insufficient ventilation of the dead space and repeated inhalation of circulating breathing medium. As a result of work being done, CO_2[19] is produced in tissues. It may not be effectively removed in the process of respiration and may accumulate in the body. This may cause an increase in blood flow, leading to increased brain exposure to oxygen and to dangerous metabolites[20] (see Chapter 3).

It seems strange that the technology of oxygen diving adopted by the US Navy does not take into account the results of the *CNSyn* risk analyses published by the authors who developed this technology (Harabin A.L., Survanshi S.S., Homer L.D., 1995; US Navy Diving Manual, 2016). Using the methods described in Chapter 4, changes were proposed in the methods applied to performing oxygen dives.

It seems that there are no major problems to propose parameters for any exposure depending on the acceptable hazard ξ of *CNSyn* occurrence. Sample calculations concerning single-depth limits are shown in Figure 5.1 and for selected multilevel profiles in Table 5.7.

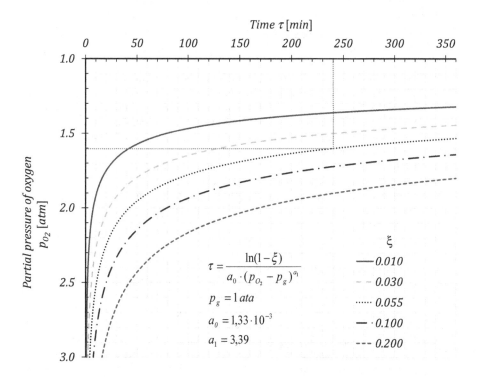

FIGURE 5.1 Oxygen partial pressure and exposures time limits in parametric relationship to ξ hazard according to equation (4.26a).

TABLE 5.7
Hazard ξ for various dive profiles according to equation (4.26a).

Dive profile time/depth	Hazard ξ according to (4.26a)
$[min]/[mH_2O]$	[%]
240/6	5.49
235/6 → 5/15	7.98
120/6 → 5/15	5.38
225/6 → 15/12	8.79
120/6 → 15/12	6.42
120/6 → 10/12	5.22
120/6 → 20/10	5.41

The developed guidelines and the experimental dives conducted on their basis were used to develop technologies of conducting oxygen dives with a permissible oxygen/*Nx* excursion (see Appendix 1).

NOTES

1 Such as magnetic barriers, acoustic barriers, etc.
2 It increases, however, the danger of detection by laser sensor or *IR* camera.
3 Diver detection sonar – *DDS*.
4 Sonar heads usually are not directed to the surface due to the interference occurring during reflection and interference of waves.
5 Target reflectivity is frequently expressed in decibels [$dB@1\ m$] as the ratio of the intensity of the wave [$W \cdot m^{-2}$] reflected from the target in the direction of the receiver at a distance of 1 *m* from its center and a flat intensity acoustic wave [$W \cdot m^{-2}$] incident on target from the receiver.
6 Therefore, it follows that for tactical reasons the minimum distance necessary to maintain secrecy of operations should be not less than 800 *m*.
7 So-called excursion/excursion to greater depths are tactical emergency activities related to the necessity to hide quickly in the depth or to avoid collisions, etc.
8 See Chapter 4.
9 See Chapter 7.2.
10 Time of breathing with oxygen.
11 This time is important from the point of view of calculation of the *CNSyn* hazard, but it should be remembered that the duration of *oxy – CCR AMPHORA SCUBA* is shorter due to the time of protective effect of the absorber (see Chapter 1).
12 This happens when dives are not too deep or too long.
13 One of such operations was described in Chapter 6.
14 With the use of other than oxygen breathing medium.
15 Without any modifications.
16 For example, air, nitrox (*Nx*), heliox (*Hx*), trimix (*Tx*), etc.
17 (18, 21)°C.

18 It should be noted that the temperature considered should apply to the diver's thermal comfort rather than directly to the average water temperature, because depending on the applied thermal protection its impact can be significantly reduced.

19 When breathing oxygen on the surface a small increase in its production occurs, but under increased pressure CO_2 production is significantly higher (Harabin A.L., Survanshi S.S., Homer L.D., 1994).

20 During descent while performing strenuous work and breathing oxygen increased cerebral blood flow was observed, although the oxygen receptors located in the carotid glomeruli experiencing the increased partial pressure lower respiratory action and slow down blood flow to the circuit.

REFERENCES

Butler FK & Thalmann ED. 1986a. Central Nervous System oxygen toxicity in closed-circuit SCUBA divers II. *Undersea Biomedical Research*, 13, 193–223.

Butler FK & Thalmann ED. 1986b. *CNS Oxygen Toxicity in Closed-circuit SCUBA Divers III*. Panama City: USN Experimental Diving Unit. Report No 5-86.

Butler FK & Thalmann ED. 1984. CNS oxygen toxicity in closed-circuit SCUBA diving. In *Proceedings of the Eight Symposium Underwater Physiology*, ed. AJ Bachrach & MM Matzen, 15–30. Bethesda: Undersea Madical Society.

Donald KW. 1992. *Oxygen and the Diver*. Harley Swan: The SPA Ltd. ISBN 1-85421-176-5.

Donald KW. 1955. Oxygen bends. *J. Appl. Physiol.*, 7, 639–644.

Harabin AL, Survanshi SS & Homer LD. 1994. *A Model for Predicting Central Nervous System Toxicity from Hyperbaric Oxygen Exposure in Man: Effects of Immersion, Exercise, and Old and New Data*. Bethesda: Naval Medical Research Institute. NMRI 94-0003; AD-A278 348.

Harabin AL, Survanshi SS & Homer LD. 1995. A model for predicting central nervous system toxicity from hyperbaric oxygen exposure in humans. *Toxicol. Appl. Pharmacol.*, 132, 19–26.

Pozański P. 2011. Współczesne zagrożenia elementów infrastruktury morskiej oraz systemy ich detekcji. *Pol. Hyperb. Res.*, 35, 7–34.

US Navy Diving Manual. 2008. The Direction of Commander: Naval Sea Systems Command. 0910-LP-106-0957.

US Navy Diving Manual. 2016. *Praca zbiorowa (revision 7)*. The Direction of Commander: Naval Sea Systems Command. SS521-AG-PRO-010 0910-LP-115-1921.

6 Oxygen/Nitrox Exposures

The *oxy – CCR/Nx – SCR AMPHORA SCUBA* diving apparatus is a construction that allows a diver to switch the supply of breathing medium from nitrox (*Nx*) to oxygen and back again. Such systems are most commonly used to support deployment of combat divers from large submarines at a maximum depth of 24 mH_2O (Figure 6.1).

A diver leaves the submarine using *Nx* in a semiclosed circuit mode *Nx – SCR AMPHORA SCUBA*, then he/she moves to the depth of transit at 6 mH_2O and switches to oxygen supplied in the closed circuit mode of *oxy – CCR AMPHORA SCUBA*.

6.1 TACTICAL CONSIDERATIONS

Tactical considerations for *oxy – Nx* dives are the same as for those described previously for oxygen[1] dives, and they were used as guidelines for planning the experimental dives. This chapter describes the possibility of performing longer and deeper trips while breathing *Nx* rather than using oxygen. However, the semiclosed circuit of *Nx* generates increased problems in keeping a combat mission clandestine because when the *Nx – SCR AMPHORA SCUBA* is in the mix configuration, part of the breathing medium must be exhausted to the surrounding water.

The previously defined radius of 800 *m*, adopted as the smallest when approaching an unreconnoitered shore because of the possible presence of onshore diver infiltration countermeasures,[2] suggests that the portion of the *Nx* dive that follows the oxygen portion, up to a certain exposure time, can be treated as not made. Therefore, no decompression is required and flying is permitted[3] at an altitude not exceeding 300 *m*.[4] As the breathing medium supply in *Nx – SCR/oxy – CCR* can be changed, the possibility of conducting *Nx* trips deeper than 15 mH_2O was considered so that the diver could hide or perform a task at a greater depth.[5] However, the change in the breathing medium required testing the purging procedure of the breathing space in the diving apparatus.

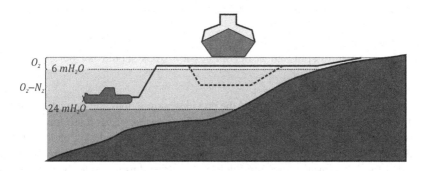

FIGURE 6.1 Typical dive profile for *Nx-SCR/oxy-CCR AMPHORA SCUBA*.

DOI: 10.1201/9781003309505-7

6.2   DECOMPRESSION PLANNING

The theoretical decompression fundamentals used here for planning nitrox exposures have been described previously and will not be discussed here again (Kłos R., 2007; Kłos R., 2011; Kłos R., 2011a).

In planning decompression for the *Nx − SCR/oxy − CCR AMPHORA SCUBA* apparatus, the *Bühlmann approach* was used with a set of *ZH − L* 16*B* tables and the *M−values* (Bühlmann A.A., 1984; Bühlmann A.A., 1995). The effects of moisture content in the inhaled and exhaled breathing medium were omitted. A balanced saturation gradient $\delta = 90\%$ has been adopted as safe. Like in the case of the *mix − SCR CRABE SCUBA* apparatus, the calculation methods described further are referred to as the AMW approach (Kłos R., 2011). An element that was not been taken into account during the calculations was flushing nitrogen from the body when oxygen was used for breathing[6] and its effect on the decompression processes.

The Bühlmann approach was applied for a comparative analysis using the program *Deco Planner 3.1.4.*[7] Saturation gradients δ[8] according to AMW are more conservative for long exposures than the usual values used in Deco Planner 3.1.4.[9]

When planning combat missions with the *Nx − SCR/oxy − CCR AMPHORA SCUBA* apparatus, attention was focused on the direct decompression profiles.

To assess the safety of experimental decompressions, the method for ultrasonic detection of the intravascular gas phase, tested in the course of investigations on the decompression for the *mix − SCR CRABE SCUBA* apparatus, was used (Kłos R., 2010; Kłos R., 2011).

6.3   ASSUMPTIONS UNDERLYING THE EXPERIMENTS

The following guidelines, derived from the initial assumptions, were adopted:

1. Normal diving procedures applicable to the ascent to surface phase required only zero decompression.
2. During an emergency extension of exposure at depth, a decompression stop at a depth of only 3 mH_2O could be authorized.
3. Maximum gradient δ according to *ZH − L* 16*B* could be changed depending on the results of diving experiments.
4. Diver's workload was simulated by swimming with fins with the preselected intensity.
5. The content of C_{O_2} in the breathing medium inhaled by the diver at the excursion depth and at decompression stops should not be less than $C_{O_2} \geq 30\%_v$.

The calculations made according to the previous assumptions are shown in Table 6.1.

It was assumed that the maximum standard *Nx* exposure τ after a minimum 45 *min* of oxygen exposure[10] $C_{O_2} \geq 70\%_v O_2$ at a depth of 3 mH_2O would not be longer than $\tau \leq 30$ *min*. The last 5 *min* of the allowed exposure is an operational emergency time intended for compensating potential delays that arise during a real dive in combat conditions.[11] Adopting the preoxygenation depth is associated with the fact that

TABLE 6.1

Table of experimental decompression for *Nx* according to AMW, assuming the composition of the circulating composition $Nx > 30\%_v$, O_2/N_2 and after the previous stay of 45 *min* at a depth of 3 *mH$_2$O* and oxygen breathing $C_{O_2} \geq 90\%_v O_2$.

Depth [mH$_2$O]→	24			21		
Bottom time	Speed of accent to first stop	Decompression stop at 3 mH$_2$O	Gradient and number of leading theoretical tissue according to AMW at bottom and 3 mH$_2$O		Decompression stop at 3 mH$_2$O	Gradient and number of leading theoretical tissue according to AMW at bottom and 3 mH$_2$O	
		Time at stop	[% of gradient:number of leading tissue]		Time at stop	[%:number of leading tissue]	
[min]	[min]	[min]			[min]		
		Basic table					
			Bottom	3 mH$_2$O		Bottom	3 mH$_2$O
10		–	28:1	45:1	–	21:1	37:1
20		–	43:1a	66:2	–	34:1a	55:2
30		–	55:2	80:3	–	43:2	67:2
40	(1–5)m · min⁻¹	–	63:3	93:3	–	50:3	78:3
50		–	–	–	–	56:3	87:4
55		–	–	–	–	58:3	92:4
		Procedures related to emergency extension of stay at depth					
			Bottom	3 mH$_2$O		Bottom	3 mH$_2$O
50		3	70:3	92:4	–	–	–
55		5	74:4	89:4	–	–	–
60		6	78:4	92:5	1	61:4	92:4
65		–	–	–	2	64:4	91:4
70		–	–	–	3	67:4	91:5

Depth [mH$_2$O]→	32		
Bottom time	Speed of accent to first stop	Decompression stop at 3 mH$_2$O	Gradient and number of leading theoretical tissue according to AMW at bottom and 3 mH$_2$O
		Time at stop	[% of gradient:number of leading tissue]
[min]	[min]	[min]	
		Basic table	
			Bottom	3 mH$_2$O
10	(1–5)m · min⁻¹	–	45:1	64:1
15		–	59:1a	81:1a

TABLE 6.1 (Continued)
Table of experimental decompression for Nx according to AMW, assuming the composition of the circulating composition Nx > 30%$_v$, O_2/N_2 and after the previous stay of 45 min at a depth of 3 mH_2O and oxygen breathing $C_{O_2} \geq 90\%_v O_2$.

Depth [mH_2O]→		32		
Bottom time	Speed of accent to first stop	Decompression stop at 3 mH_2O	Gradient and number of leading theoretical tissue according to AMW at bottom and 3 mH_2O	
		Time at stop	[% of gradient:number of leading tissue]	
[min]	[min]	[min]		
		Procedures related to emergency extension of stay at depth		
			Bottom	3 mH_2O
20	$(1-5)m \cdot min^{-1}$	1	69:2	87:2
25		2	79:2	92:3
30		4	85:2	92:4
35		6	93:3	95:4

Maximum gradient according to ZHL 16B will not be greater than ca. 90%$_v$.

Oxygen content C_{O_2} at depth 3 mH_2O cannot be smaller than $C_{O_2} \geq 90\%_v$.

Oxygen content C_{O_2} at depth (21, 24) mH_2O and during the decompression cannot be smaller than $C_{O_2} \geq 30\%_v$.

staying at depths shallower than 3 mH_2O is difficult during a maritime transit and in large lakes due to the impact of waves on the diver. Transit is performed mostly within the depth range $H \in (4, 5)$ mH_2O. Hence and following the previous tactical analyses, using the worst-case scenario, the preoxygenation was adopted as a 45 min long dive at a depth of 3 mH_2O.

The use of the conservative algorithm $ZH - L$ 16B increases the diving safety without any significant negative impact on the properties of the technology proposed for tactical diving.

The selected operating time of 30 min results from the tactical considerations[12] and the average consumption of Nx during a dive (AQUA LUNG, 2004) (see Chapter 1).

For this reason, 30 min should not be exceeded in planning and executing combat dives.

Nx 43%$_v O_2/N_2$ was selected from Table 1.4 as a premix that could be used for diving to a depth of 32 mH_2O. In this case, the safe exposure is 28 min[13] (AQUA LUNG, 2004).

6.4 CHANGES IN DEPTH

During the first training period and Nx dives not preceded by preoxygenation, the descent to 10 mH_2O was planned to be at a rate not exceeding 8 $mH_2O \cdot min^{-1}$. After reaching a depth of 10 mH_2O the descent rate could be faster, but was not to exceed 20 $mH_2O \cdot min^{-1}$.

Once divers mastered the diving using $oxy - CCR/Nx - SCR\ AMPHORA\ SCUBA$, they proceeded to the exposure from the transit depth after switching the apparatus from $oxy - CCR\ AMPHORA\ SCUBA$ mode to $Nx - SCR\ AMPHORA\ SCUBA$ at the maximum possible speed, which is limited only by the necessity to fill up the breathing space in the apparatus through bypass valve 12 in Figure 1.3.

6.5 DECOMPRESSION

In this project Nx dives were only complementary to the oxygen dives, increasing opportunities to escape into depth using Nx after preoxygenation associated with a transit phase. It was assumed that preference would be given to the zero decompression procedures. Table 6.2 presents a theoretical decompression table adopted for the purposes of this study. It is divided into a basic table and emergency procedures.[14] The emergency procedures were not tested in the project.[15]

TABLE 6.2
AMW experimental decompression table for circulating $Nx > 30\%_v\ O_2/N_2$ and after a previous stay of 45 min at a depth of 3 mH_2O, breathing oxygen in configuration $oxy - CCR\ AMPHORA\ SCUBA$.

Depth	Time at bottom	Speed of ascent to first stop	Decompression stops [mH_2O]		Total decompression time
			6	3	
			Time at stop		
[mH_2O]	[min]	[$m \cdot min^{-1}$]	[min]		[min]
21	55	[1, 5]	–	–	–
		Procedures regarding emergency extension of stay			
	60	[1, 5]	–	1	1
	67		–	2	2
	70		–	3	3
24	40	[1, 5]	–	–	–
		Procedures regarding emergency exposure extension			
	55	[1, 5]	–	3	3
	60		–	4	4
	65		–	5	5
32	15	[1, 5]	–	–	–
		Procedures regarding emergency exposure extension			
	20	[1, 5]	–	1	1
	25		–	2	2
	30		–	4	4

Minimal transit time between stop at 3 mH_2O and surface is 30 s.

6.6   PURGING PROCEDURES

One of the subjects of the study was the flushing/purging procedures of the apparatus breathing space when there was a change of the breathing medium from Nx
to oxygen. The start of a dive was preceded by purging the breathing space with
oxygen on the surface. This procedure was repeated three times (see Chapter 7).
Immersion to the depth of the trip was not preceded by rinsing the breathing space,
only switching to power Nx and opening the relief valve.[16] The breathing medium
was added to the circuit only as needed. After the scheduled trip time, there was
a change to the oxygen supply and the relief valve was closed. The breathing bag
was then emptied by exhausting the breathing medium, in a controlled manner,[17]
into water. This was followed by a slow ascent with normal breathing. During the
tests changes in the procedures of purging the breathing space were allowed. These
changes were presented in the description of the test results.

6.7   ARRIVAL TIME

According to the tactical situation guidelines after a nitrox trip, the diver was to
ascend to the surface. It was assumed that it would be necessary to repeat such an
operation, as after the escape into deep water divers could easily lose the required
direction of swim.

Such an operation will also be needed if the basic objective of the mission is
abandoned and there is a need for the diver to change to a contingency or emergency
direction.

Ascent to the surface immediately after the trip is more decompression loaded
than ascent only to a depth of transit. Thus, in experimental dives a decision was
made to enact this kind of combat exercise scenario, which meant applying the
method of the worst-case scenario to test the system.

Ascent time should be calculated in such a way that the emission of expanding
breathing medium into the water, which always accompanies ascent, is minimal. In
such a case maintaining secrecy of the mission may be difficult. Therefore, the ascent
technique must be permanently improved. During ascent two adverse events related
to an increase in volume of breathing medium in the circuit occur. The first one produces the already mentioned effect of increased emission of gases into the water. The
other is the effect of additional lift, which could in an extreme case lead to the loss of
buoyancy control and ejection of the diver, in an uncontrolled manner, to the surface.

6.8   SECOND DESCENT

Before the second descent the breathing space of apparatus in $oxy - CCR\,AMPHORA$
$SCUBA$ configuration was purged with oxygen, so its concentration in the breathing
medium immediately increased. Such an action under normal operating conditions
is not required, and intensive purging may even be harmful for protection against
$CNSyn$. Currently, the aim is to ensure that during a dive the diver breathes nitrox in
which O_2 content does not exceed $85\%_v O_2/N_2$ (Harabin A.L., Survanshi S.S., Homer
L.D., 1994; Walters K.C., Gould M.T., Bachrach E.A., Butler F. K., 2000). During

the experimental dives, however, efforts were made to increase the concentration of oxygen as soon as possible to find out if there were any oxygen toxicity problems causing *CNSyn* symptoms.

6.9 SUMMARY

The analyses conducted and experimental dives gave way to developing an original technology of oxygen-nitrox diving with the *oxy – CCR/Nx – SCR AMPHORA SCUBA* diving apparatus (see Appendix 1).

This chapter describes the possibility of making a no-decompression *Nx* trip as an extension of the technology of oxygen transit with excursion diving.[18]

During this study, the methods of purging the breathing space of a diving apparatus were improved and tested. Decompression safety was also tested.

The proposed technology of diving covers the most likely tactical scenarios.

This technology should be further supplemented by important elements such as safety of helicopter transport at an altitude of above 300 *m*, the impact of the combat-related effort[19] on the safety of underwater escape, the impact of underwater transport on combat capabilities and so on. Not without significance is finding out changes in individual resistance to decompression or exposures to oxygen high partial pressure during adaptation training.

NOTES

1 Since *Nx* trips are only the extension of oxygen diving technology.
2 For tactical reasons therefore, the minimum distance for operations secrecy should be 800 *m*.
3 Due to preoxygenation.
4 The ability to recover reconnaissance/special groups/sections by air using a helicopter operating at an altitude higher than 300 *m* following the rapid rebound after completing an oxygen mission should be confirmed by testing.
5 Problems connected with disembarking the divers from the submarine were not checked at the time of this project.
6 Preoxygenation.
7 However, our own method of calculating decompression tables should be used due to the fact that Deco Planner 3.1.4 does not allow the analysis of the profiles considered by it as posing a risk of *DCS*.
8 Gradients larger than $\delta = 90\%$ of maximum value were initially considered here to be potentially dangerous in view of the risk of *DCS* unless studies proved otherwise.
9 Suggested by Deco Planner 3.1.4, the varying permeability model (VPM) method for the generation of the tables was not used here because it is more conservative.
10 This assumption results from the tactical scenario described in Chapter 5.
11 This allows for maintaining standard zero decompression in the case of a short delay.
12 Generally, there is no need for a longer time to perform the task or escape to depth between 6 mH_2O and 24 mH_2O.
13 36 *min* without taking into account the reserve supply of 50 *atn* of respiratory medium – Table 1.5.
14 If emergency procedures could be generated in the course of theoretical calculations for each block representing various decompression assumptions.

15 They act as boundary procedures that should not be exceeded and could only be used in an emergency.
16 Simulation of a quick escape in combat.
17 Emissions of large bubble-dense groups should be avoided to maintain the operation's secrecy.
18 such technology has not been described so far.
19 Attack, engagement, etc.

REFERENCES

AQUA LUNG. 2004. *AMPHORA Oxygen and Mix Gas Diving Equipment – Operating and Maintenance Manual*. Nice: La Spirotechnique I.C. AMPHORA Réf. A658.

Bühlmann AA. 1984. *Decompression-Decompression Sickness*. Berlin: Springer-Verlag. ISBN 3-540-13308-9; ISBN 0-387-13308-9.

Bühlmann AA. 1995. *Tauchmedizin*. Berlin: Springer-Verlag. ISBN 3-540-58970-8.

Harabin AL, Survanshi SS & Homer LD. 1994. *A Model for Predicting Central Nervous System Toxicity from Hyperbaric Oxygen Exposure in Man: Effects of Immersion, Exercise, and Old and New Data*. Bethesda: Naval Medical Research Institute. NMRI 94-0003; AD-A278 348.

Kłos R. 2010. Detekcja śródnaczyniowej wolnej fazy gazowej. *Pol. Hyperb. Res.*, 32, 15–30.

Kłos R. 2011a. Jednotkankowy model dekompresji ograniczonej procesem dyfuzji. *Pol. Hyperb. Res.*, 35, 69–94.

Kłos R. 2011. *Możliwości doboru dekompresji dla aparatu nurkowego typu CRABE – założenia do nurkowań standardowych i eksperymentalnych*. Gdynia: Polskie Towarzystwo Medycyny i Techniki Hiperbarycznej. ISBN 978-83-924989-4-0.

Kłos R. 2007. Niektóre problemy związane z wyborem sposobu dekompresji. *Pol. Hyperb. Res.*, 18, 33.

Walters KC, Gould MT, Bachrach EA & Butler FK. 2000. Screening for oxygen sensitivity in US Navy combat swimmers. *Undersea Hyper. Med.*, 27, 21–26.

7 Experimental Diving

7.1 GENERAL GUIDELINES FOR EXPERIMENTS

Experiments were carried out in diving complex *DGKN-120* at the Naval Academy in Gdynia (Figure 7.1).

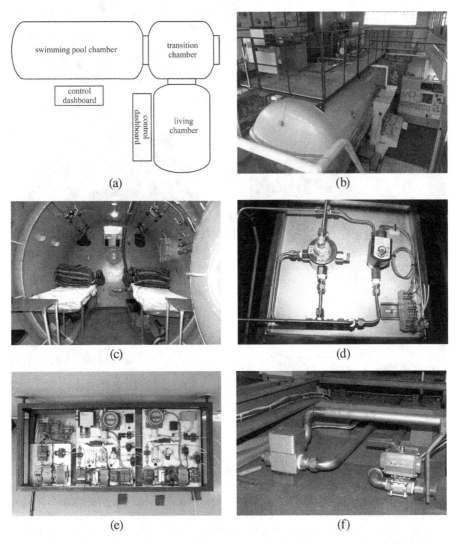

FIGURE 7.1 Experimental Deepwater Diving Complex DGKN-120: (a) diagram of chambers making the complex, (b) complex DGKN-120 viewed from the top, (c) living quarters

DOI: 10.1201/9781003309505-8

FIGURE 7.1 (*Continued*)
viewed from the transit chamber, (d) nozzle system fitted to the automatic oxygen dosing system, (e) system for automatic measuring atmospheric composition inside the main and transit chamber, (f) actuator of a system for automatic maintaining the pressure and controlling the process of decompression.

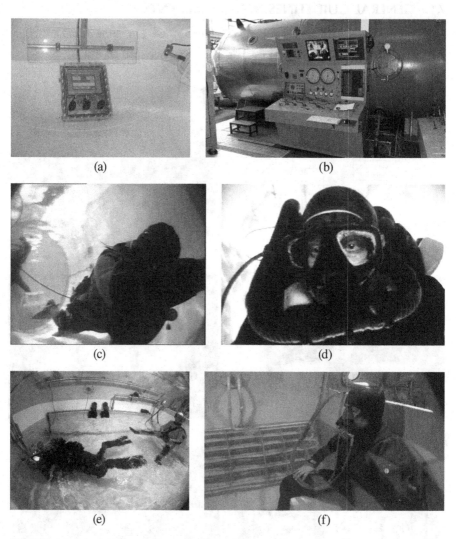

FIGURE 7.2 Experimental Deepwater Diving Complex DGKN-120: (a) computer with data for the diver, (b) diver at an ergometer post looking at the indicators and data that appear on the computer screen, (c) side view from the camera used to correct the primary diver's position in relation to the ergometer, (d) camera view showing the face of the primary diver, (e) system designed to control and monitor the experiment, (f) standby diver during decompression using *BIBS*.[2]

7.1.1 ORGANIZATION OF EXPERIMENTS

A diver under a workload in the swimming pool of diving complex DGKN-120 was accompanied by a standby diver (Figure 7.2f). During the experimental exposure the diver was in the water breathing from the diving apparatus *oxy – CCR/mix – SCR AMPHORA SCUBA* (Figures 7.2c–7.2e). The exercise system included a computer capable of displaying: commands, some diving parameters, hydrostatic force and oxygen content in the breathing medium[1] (Figure 7.2a). There were three computer buttons a diver could use for quick communication.

To communicate with the diver, a wired communication system could be used or a loudspeaker could be located inside the swimming pool to give the diver direct commands by the water environment. In the computer box, a camera to observe the diver's face is installed (Figure 7.2d). The side camera shows the accuracy of the diver's position under water (Figure 7.2c). Inside the complex, cameras show the general view inside the chambers (Figure 7.2b).

Before a dive the standby diver connected the breathing medium transfer assembly fitted on the inhaling hose of the test diver's apparatus[3] using a quick coupling. By applying this method, the breathing medium inhaled by the diver was collected for analysis. The maximum sampling flow of the inhaled breathing medium was $V < 1 \ cm^3 \cdot min^{-1}$. This flow was directed to the SERVOMEX Pm1111E701 Oxygen Transducer[4] paramagnetic oxygen analyzer. The analyzer is an element of the automatic system for measuring breathing medium composition inhaled by the test diver. The system similar to that installed in each chamber of the DGKN-120 complex (Figure 7.1e). The complex is equipped with devices for maintaining constant pressure and making automatic pressure changes in accordance with the planned decompression by actuator shown in Figure 7.1f. In the atmosphere inside the hyperbaric chamber, the constant partial pressure of oxygen was maintained by actuator shown in Figure 7.1d. Data provided by the devices measuring composition of breathing medium, pressure level, decompression, oxygen partial pressure, and tension[5] were monitored, displayed and saved[6] on one of the randomly selected computers making up the measuring network of diving complex DGKN-120. The actuators were controlled and operated by the same selected computer. In addition, the data were stored on the main computer server. All the systems were controlled using integrated software that can operate on a selected computer connected[7] to the Ethernet measuring network (Figure 7.2b).

While staying at depth, the test diver performed work by simulating swimming with fins simultaneously pushing a vertical plate connected to an NK-SS 300N tensiometer made by Keli Electric Manufacturing (Ning 80) Co., Ltd. (Figure 7.2a). The standby diver's only task was to provide emergency assistance to the test diver. The standby diver during experimental diving could breathe the gas from the atmosphere of the chamber or *Nx*/oxygen through the BIBS equipped with half masks. The exhaled gas was exhausted through the exhaust hose outside the chamber using the system designed to maintain the necessary pressure difference. The length of the hoses allowed the standby diver to move around the whole chamber (Figure 7.2f). If the standby diver breathed the air from the atmosphere of the chamber, very often he/she had to be replaced by another standby diver before the start of decompression because his/her safe decompression schedule would be slower than that of the test diver due to the difference in oxygen content in the inhaled breathing medium.[8] The standby divers can be substituted under pressure, as all chambers in complex DGKN-120 are connected, but they may also be operated independently even when different pressures are maintained in them[9] (Figure 7.1a). After substituting the standby divers, the first standby diver undergoes decompression in the living chamber of complex DGKN-120 with the option of oxygen decompression (Figure 7.1c). The second standby diver undergoes decompression together with the test diver. The diving supervisor must choose the right moment for the substitution so that the procedure is safe and convenient as far as the experiment is concerned.

During decompression the test diver did not perform any work or performed it with the outlined intensity. If the physician supervising the procedure deemed it necessary, after the end of decompression the test diver was moved quickly to the living chamber, where he/she could be subjected to a test for the presence of intravascular free gas phase[10] (Kłos R., 2011; Kłos R., 2020). The time interval between the end of the dive and the start of the test should be no longer than 7 *min*.[11] If the doctor allowed, after the measurement the test diver took a 5 *min* long warm shower to provoke skin symptoms of *DCS*. Then the diver went back into the living chamber, where again he/she was tested for the presence of intra-vascular free gas phase.

7.1.2 Safety of Decompression

When the test diver was under water, the standby diver was in close proximity and standby ready to provide emergency assistance in case *CNSyn* symptoms occurred.

After the end of exposure, the diver was tested using a device for detecting intravascular free gas phase (Kłos R., 2011; Kłos R., 2020). The minimum observation time was 2–3 *h*[12] after the end of decompression, provided that the signal from the free gas phase was not noticed. If such a signal was recorded, the observation time was extended by (1, 2) *h* from the moment the signal coming from the free gas phase disappeared. During the tests, the right and left subclavian vein and the right atrium of the heart were monitored. Exceeding the value of grad *III* for the atrium region or grad *II+* for subclavian veins were the reason to start medical treatment (Kłos R., 2011; Kłos R., 2020). For example, such treatment might consist of normobaric flushing with oxygen, hyperbaric flushing with oxygen, according to US Navy Treatment Table *TT 5 USN* etc. (US Navy Diving Manual, 2016).

Symptoms of skin and/or joints and muscle *DCS* was an indication to immediately start medical treatment according to the Table *TT 6 USN* with its possible extension or moving to the procedure in *TT 6A USN*. All clinical signs consistent with severe forms of *DCS*[13] were a signal for the immediate initiation of medical treatment according to Table *TT 6A USN*. All decisions to undertake medical treatment at the onset of sickness symptoms were taken by a diving service physician, who then took responsibility for managing the treatment. The medical treatment procedures will not be cited here (US Navy Diving Manual, 2016).

It was permissible to provoke skin symptoms of *DCS* by taking a 5 *min* long hot shower immediately after decompression. The diver could take a shower only when appropriate assistance was available and a diving service officer/diving medical service doctor was informed about such an intention.

In the event of asymptomatic intravascular free gas phase, the decision to undertake medical treatment were made by the project manager following mandatory consultation with the medical team. The code of ethics applicable to the Naval Academy was previously published and will not be quoted here (Kłos R., 2011).

The minimum rest time after completing decompression was 12 *h*. Repetitive dives in this cycle of tests were banned. The experimental divers were not exposed to hyperbaric pressure more than three times in a six-day working week.[14] This applied only to participation in the tests.[15] With consent of the doctor the number of exposures could be increased, but to no more than six times in a six-day working week.

7.1.3 Workload

Assuming that frontal resistance R_{cz} generated by the diver together with the diving apparatus can be in the first approximation replaced by the model of a flat square plate with an area of $S_{cz} \cong 0.216\,m^2$, it can be roughly estimated as[16] $R_{cz} \equiv P \cong C_p \cdot \frac{\rho \cdot v^2}{2} \cdot S_{cz}$, where C_p –dimensionless coefficient of resistance adopted here as dimensionless value $C_p = 1.1$ for a square flat plate,[17] ρ – density of water[18] in $[kg \cdot m^{-3}]$, v – speed with which diver moves in $[m \cdot s^{-1}]$.

Hence, one can estimate the swimming pace v as a function of the measured pressure exerted by the diver on the square plate combined with tensiometer R_{cz} (Figure 7.3).

The calculation results are consistent with the feelings of the divers and the tests, which involved relating the heart rate to the physical exercise done when swimming over a distance of 200 *m* with a precisely controlled pace, carried out at the model pool of the *Ship Design and Research Centre* in Gdansk (Figure 7.4).

7.1.4 Stages of Experiments

In the first stage of the tests, the focus was placed on building teamwork capabilities and determining the training level of the experimental divers.

After testing the realizability of the combat scenarios and adopting amendments, studies on the decompression problems were commenced.

Their goal was to determine the maximum allowable exposure times for excursions to depths of up to 24 *mH₂O* using *Nx – SCR AMPHORA SCUBA* in *Nx* mode and no decompression after the previous 45 *min* oxygen exposure at 3 *mH₂O* using *oxy – CCR AMPHORA SCUBA* in *oxy* mode – preoxygenation.[19] In a separate stage of the study, random oxygen toxicity risk tests were carried out.

7.2 EXPERIMENTAL OXYGEN EXPOSURES

The divers employed *CCR AMPHORA SCUBA* in *oxy* mode for the oxygen dives, which lasted about 1.5 *h* at a transit depth of 6 *mH₂O* with one or two excursions in compliance and noncompliance with the US Navy excursion limits.

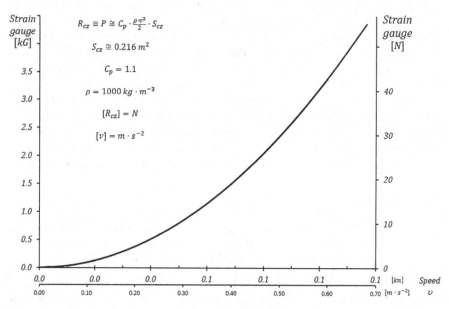

FIGURE 7.3 Calculations of resistance effect on speed of swimming that the diver has to overcome.

FIGURE 7.4 Measurements of physical exercise done when swimming 200 *m* distance at a controlled pace at the pool of the Ship Design and Research Centre in Gdansk: (a) view of the pool, (b) a bridge over the pool that can move exactly with the preset speed, (c) preparing to dive, (d) diver following the light marker attached to the moving bridge.

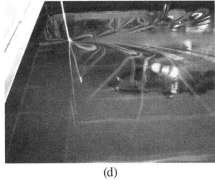

(c)   (d)

FIGURE 7.4 (Continued)

7.2.1 CONTROLLING PHYSICAL EXERCISE

The average simulated load was ranged $(2; 2.5)kg$. Descent pace to an excursion depth was within $(10, 20)mH_2O \cdot min^{-1}$. After 45 *min* of simulated swimming, a descent was made without flushing the breathing space. It corresponded to a rapid escape into depths at a speed[20] ranged $(10, 20)mH_2O \cdot min^{-1}$ to one of the two depths $\{12; 15\}mH_2O$ the workload measured with a tensiometer corresponding to about 3 kg[21] (Figure 7.3). Shortages of breathing medium were supplemented by a bypass valve[22] (marked as 12 in Figure 1.2) only when it was absolutely necessary. After reaching the planned depth of excursion the load was reduced to $(2.0, 2.5)kg$. Before the ascent from the excursion depth, the breathing bag had to be emptied by releasing the breathing medium into water in a controlled way.[23] Then ascent to a transit depth had to be made with a speed ranged $(1, 3)mH_2O \cdot min^{-1}$ and a load, being measured with a tensiometer, which corresponded to approximately 2 kg. The whole procedure had to be carefully planned and carried out in a controlled manner as the total exposure time, the excursion and the ascent to the depth of transit could not exceed the total time assumed for the excursion (Appendix 1 Figure A1.1).

After returning to the transit depth of 6 mH_2O the dive had to be continued with a load ranged of $(2.0, 2.5)kg$, measured with a tensiometer until the end of the scheduled time of the mission.

The diving apparatus was not purged either before or during the dive in accordance with the guidelines of the US Navy.[24]

7.2.2 TRANSIT PROCEDURE

Seven experimental dives were performed in the course of testing the standard oxygen exposures.

During each of the dives, a single allowable excursion to a depth of 12 mH_2O or 15 mH_2O was taken (see Chapter 5). Figure 7.5 presents the results of three experimental dives with allowed single excursions to 12 mH_2O, and Figure 7.6 shows the results of four experimental dives with allowed single excursions to a depth of 15 mH_2O.

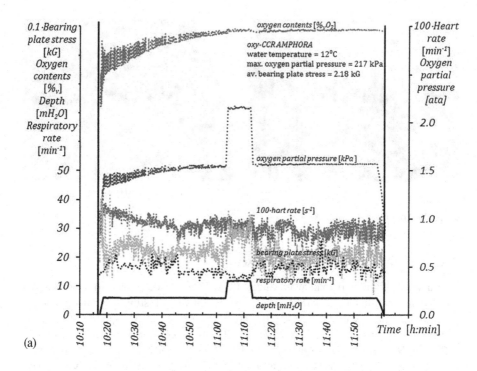

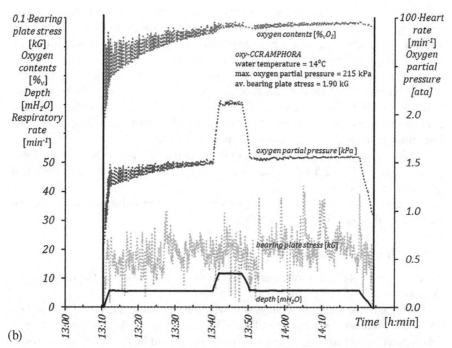

FIGURE 7.5 Results of oxygen dives with the allowed oxygen excursion on 12 mH_2O.

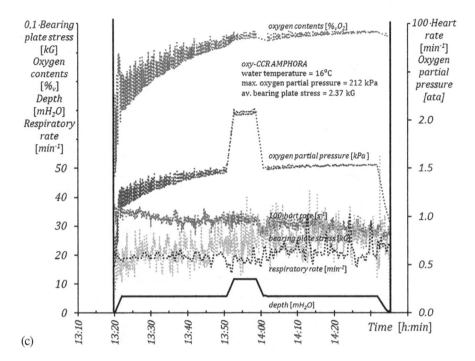

(c)

FIGURE 7.5 (Continued)

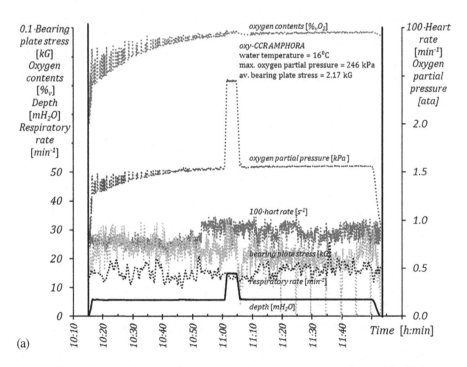

(a)

FIGURE 7.6 Results of oxygen dives with the allowed oxygen excursion on 15 mH_2O.

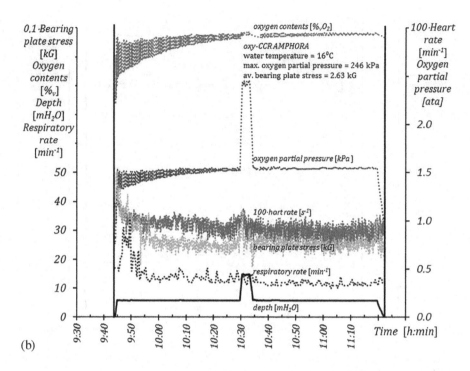

(b)

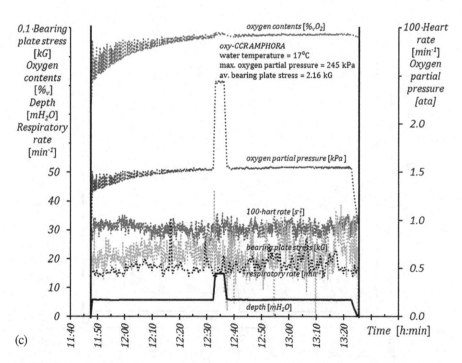

(c)

FIGURE 7.6 (Continued)

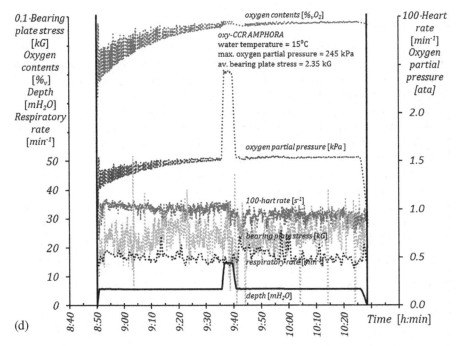

(d)

FIGURE 7.6 (Continued)

The excursions were taken after 45 *min* transit at a depth of 6 *mH₂O*. The dive was completed within the total time of approximately 1.5 *h* at a depth of 6 *mH₂O*. As it follows from the analysis in Chapter 5, combat diving operations using oxygen should take into consideration swimming with, at most, an average speed. For this reason the divers were instructed to maintain such intensity of swimming.

The average load for three dives using the transit with excursion to a depth of 12 *mH₂O* expressed by the pressure F exerted on the horizontal plate connected to tensiometer was $F \in \{2.18; 1.90; 2.37\}$ *kg* respectively. According to Figure 7.3, it was approximately taken that the recorded average pressures F correspond to the average speeds on the distances covered by the swimmers expressed in knots $v \in \{0.8_2; 0.7_7; 0.8_6\} kn$.

The divers' heart rates during exposure stabilized within the range[25] $HR \in (92, 94)$ *min⁻¹*.

The tests were performed in water at a temperature t, which was $t \in \{12; 14; 16\}°C$ respectively.

After maintaining $(20, 25)$ *min* the transit depth the oxygen content in the circuit exceeded $C_{O_2} > 96\%_v$. The maximum oxygen partial pressure P_{O_2} during the excursion was $p_{O_2} \in \{217; 215; 212\} kPa$.

The average load for four dives using the transit with excursion to a depth of 15 *mH₂O* expressed in the pressure F exerted on the horizontal plate connected to tensiometer was $F \in \{2.17; 2.63; 2.16; 2.35\}$ *kg* respectively. According to Figure 7.3, it was approximately recorded that the average observed pressures F correspond to the average speeds of distances covered by the swimmers, expressed in knots $v \in \{0.8_2;$ $0.9_1; 0.8_2; 0.8_6\}$ kn.

The divers' heart rates stabilized within the range of HR $\in$ (84, 97)min^{-1}. During the exposure in Figure 7.6a, the heart rate initially stabilized at approximately $HR = 76$ min^{-1}, before an excursion was made it went up to approximately $HR = 90$ min^{-1} and then stabilized at this level.

Tests were performed in water at $t \in \{16; 16; 17; 15\}$ °C respectively.

After maintaining the approximately (20, 25) min transit depth, the content of oxygen in the circuit exceeded $C_{O_2} > 96\%_v$. The maximum oxygen partial pressure p_{O_2} during the excursion was $p_{O_2} \in \{246; 246; 245; 245\}kPa$.

During the experimental dives, no symptoms of oxygen toxicity were observed and the divers completed the tests in good condition.

The final value of the cumulative hazard function ξ of onset of *CNSyn* symptoms according to the algebraic semi-empirical mathematical model (4.26a) for a transit to a depth of 6 mH_2O with allowed 10 min excursion to a depth of 12 mH_2O is:

$$\xi = \xi\left(t = 45min; p_{O_2} = 160\,kPa\right) + \xi\left(t = 10\,min; p_{O_2} = 220\,kPa\right) + \xi\left(t = 45\,min; p_{O_2} = 220\,kPa\right) + \xi\left(t = 45\,min; p_{O_2} = 160\,kPa\right) \cong 0.045 \triangleq 4.5\% \quad \text{(Figure 7.5)}.$$

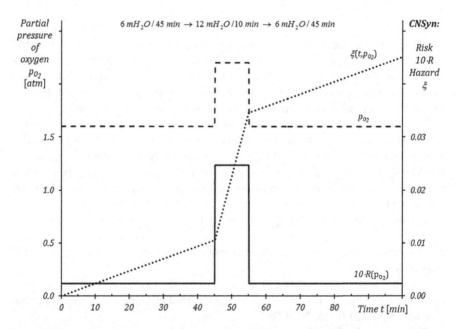

FIGURE 7.7 Dependence of risk R and the cumulative hazard function ξ of onset of *CNSyn* symptoms on time t according to the semi-empirical algebraic mathematical model (4.26a) for a dive using transit with an allowed 10 min excursion to 12 mH_2O (Harabin A.L., Survanshi S.S., Homer L.D., 1995).

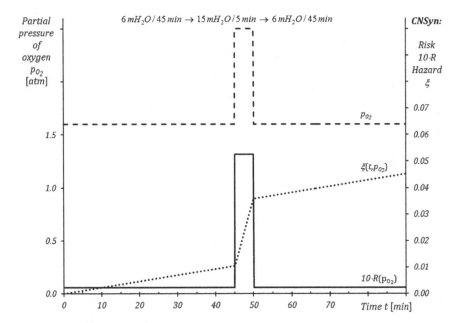

FIGURE 7.8 Dependence of risk R and the cumulative hazard function ξ of onset of *CNSyn* symptoms from time t according to the semi-empirical algebraic mathematical model (4.26a) for a dive with allowed 5 *min* excursion to 15 mH_2O (Harabin A.L., Survanshi S.S., Homer L.D., 1995).

The final value of the cumulative hazard function ξ of onset of *CNSyn* symptoms according to the algebraic semi-empirical mathematical model (4.26a) for transit to a depth of 6 mH_2O with allowed 5 *min* excursion to a depth of 15 mH_2O is:

$$\xi = \xi\left(t = 45\,min; p_{O_2} = 160\,kPa\right) + \xi\left(t = 5; p_{O_2} = 250\,kPa\right) + \xi\left(t = 45\,min; p_{O_2} = 160\,kPa\right) \cong 0.047 \triangleq 4.7\%$$

(Figure 7.6. Hazard below $\xi \le 5\%$ is acceptable for combat diving.

7.2.3 Exceeding Time Allowed for Excursion

Six dives were conducted in which the time of excursion exceeded the allowed limits (see Chapter 5). The dives were conducted at a depth of 6 mH_2O with a single excursion taken after 45 *min* transit to a depth of $H \in \{12; 12; 12; 12; 15; 15\}$ mH_2O exceeding the allowed time limit. Then the dive was completed within the total time of approximately 1.5 h at a transit depth of 6 mH_2O (Figures 7.9–7.10).

The average load for these dives expressed in pressure F exerted on the horizontal plate connected to tensiometer was at the level of $F \in \{2.01; 2.31; 1.98; 2.02; 2.27; 2.57\}$ kg respectively. According to Figure 7.3 it can be assumed that the average observed pressures F correspond to the average speeds on distances covered by the swimmers, expressed in knots $v \in \{0.7_9; 0.8_5; 0.7_9; 0.8_0; 0.8_4; 0.9_0\}$ kn.

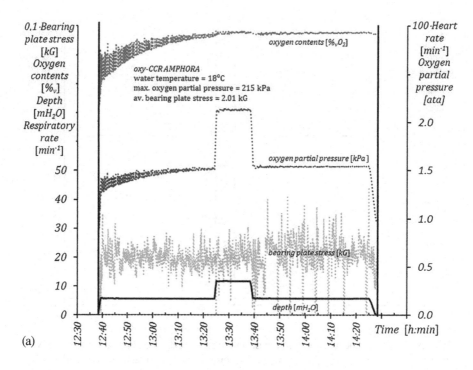

(a)

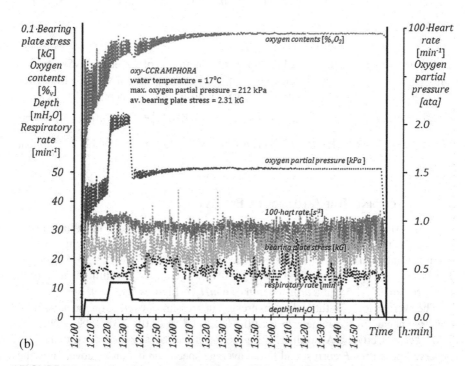

(b)

FIGURE 7.9   Results of oxygen dives with a prohibited 12 mH_2O oxygen excursion.

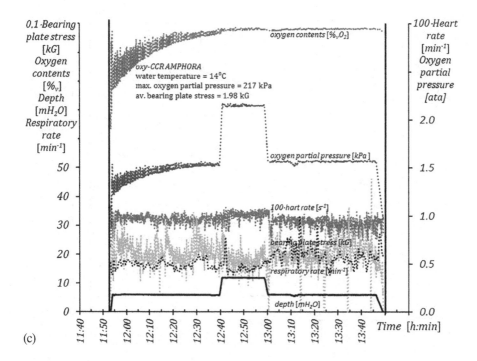

(c)

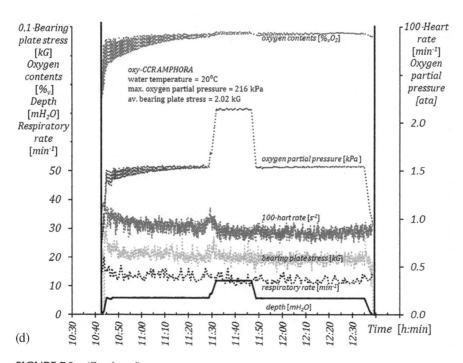

(d)

FIGURE 7.9 (Continued)

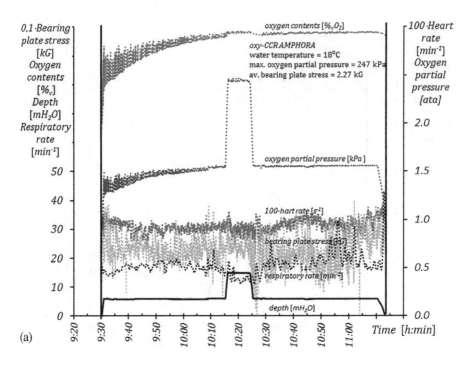

(a)

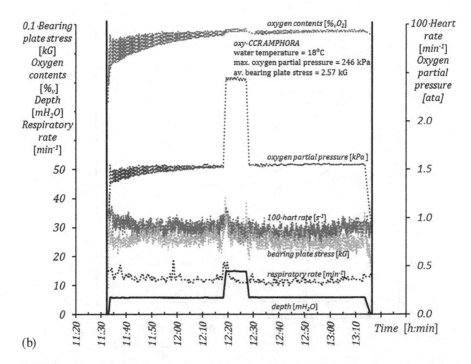

(b)

FIGURE 7.10 Results of oxygen dives with a prohibited 15 mH_2O oxygen excursion.

Divers' heart rates during the exposures stabilized within the range of[26] HR ∈ (89, 96) *min*⁻¹.

The tests were performed in water at temperature $t \in$ {18; 17; 14; 20; 18; 18} °C respectively. After maintaining an approximately (20, 25) *min* transit depth the content of oxygen in the circuit exceeded $C_{O_2} > 96\%_v$. The maximum partial pressure during excursions performed sequentially at a depth $H \in$ {12; 12; 12; 12; 15; 15} *mH₂O* with the corresponding time $\tau \in$ {15; 15; 20; 20; 20; 10; 10} *min* amounted to $p_{O_2} \in \{215; 212; 217; 216; 247; 246\} kPa$.

During the dives there were no signs of oxygen toxicity and the divers completed the tests in good condition.

The final value of the cumulative hazard function ξ of onset of *CNSyn* symptoms according to the algebraic semi-empirical mathematical model (4.26a) for transit to a depth of 6 *mH₂O* with unallowable 20 *min* excursion to a depth of 12 *mH₂O* is:

$$\xi = \xi\left(t = 45\,min;\, p_{O_2} = 160\,kPa\right) + \xi\left(t = 20min;\, p_{O_2} = 220\,kPa\right) + \xi\left(t = 45\,min;\, p_{O_2} = 160\,kPa\right) \cong 0.069 \triangleq 6.9\%$$ (Figure 7.11).

Usually it is assumed that for typical combat missions the permissible hazard is $\xi \leq 5\%$, hence the hazard of approximately $\xi \cong 7\%$ is usually not approved for combat dives.

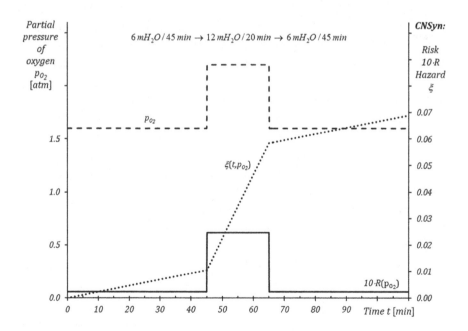

FIGURE 7.11 Dependence of risk R and the cumulative hazard function ξ of onset of *CNSyn* symptoms from time t according to the semi-empirical algebraic mathematical model (4.26a) for a transit dive at 6 *mH₂O* with unallowable excursion to 12 *mH₂O* (Harabin A.L., Survanshi S.S., Homer L.D., 1995).

The final value of the cumulative hazard function ξ of onset of *CNSyn* symptoms according to the algebraic semi-empirical mathematical model (4.26a) for transit to a depth of 6 mH_2O with unallowable 10 *min* excursion to a depth of 15 mH_2O is:

$$\xi = \xi\left(t = 45\,min, p_{O_2} = 160\,kPa\right) + \xi\left(t = 10\,min, p_{O_2} = 250\,kPa\right) + \xi\left(t = 45\,min, p_{O_2} = 160\,kPa\right) \cong 0.072 \triangleq 7.2\%$$ (Figure 7.12). Hazard at a level of $\xi \leq 7\%$ is usually not accepted[27] for combat dives.

7.2.4 DOUBLE EXCURSIONS

Two dives were conducted in which time of the excursion limits were exceeded at a depth of 12 mH_2O (see Chapter 5).

The dives started at a transit depth of 6 mH_2O. After 45 *min* of transit a 15 *min* excursion was taken to a depth $H = 12$ mH_2O. After returning to the transit depth and remaining for the next 45 *min* at a depth of 6 mH_2O, another 15 *min* excursion was taken completing the diving process. The time of the second excursion included decompression to the surface. The whole process lasted 2 *h* (Figure 7.13).

The average load for these dives, expressed in the pressure F exerted on the horizontal plate connected to tensiometer, was at a level of $F \in \{2.37; 2.26\}$ *kg*

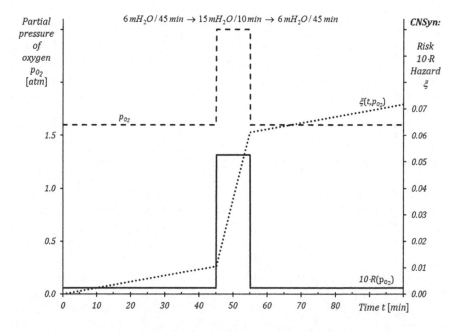

FIGURE 7.12 Dependence of risk R and the cumulative hazard function ξ of onset of *CNSyn* symptoms from time t according to the semi-empirical algebraic mathematical model (4.26a) for a transit dive to 6 mH_2O with unallowable excursion to 15 mH_2O (Harabin A.L., Survanshi S.S., Homer L.D., 1995).

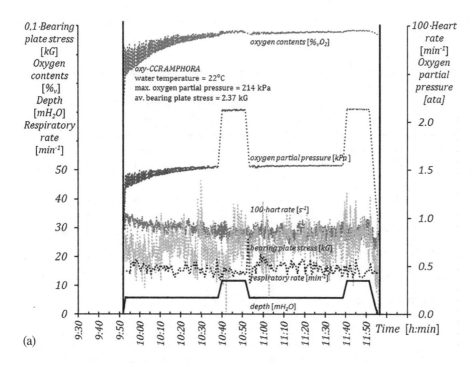

(a)

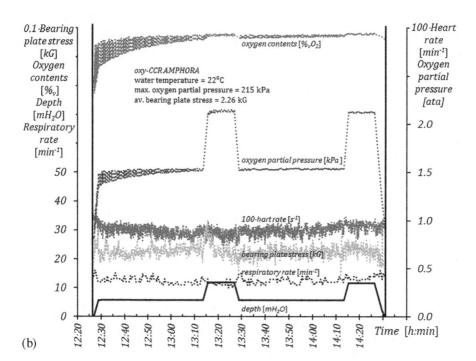

(b)

FIGURE 7.13   Results of oxygen dives with a double trip to a depth of 12 mH_2O.

respectively. According to Figure 7.3 it can be assumed that the average observed pressures F correspond to the average speeds of the distances covered by the divers, expressed in knots $v \in \{0.8_6; 0.8_4\}$ kn.

Divers' heart rates during the exposure stabilized within $HR \in (88, 90)$ min^{-1}.

Both dives were carried out in water at $t \cong 22°C$.

After maintaining an approximately $(20, 25)min$ transit depth the oxygen content in the circuit exceeded $C_{O_2} > 96\%_v$. The maximum partial pressure during the excursions at a depth H = 12 mH_2O and excursion time $\tau = 15$ min were in the first dive $p_{O_2} \in \{214; 214\} kPa$ and in the second $p_{O_2} \in \{215; 215\} kPa$.

During the dives there were no signs of oxygen toxicity and divers completed the tests in good condition.

The final value of the cumulative hazard function ξ of onset of *CNSyn* symptoms according to the algebraic semi-empirical mathematical model (4.26a) for transit with excursion dives to a depth of 6 mH_2O with unallowable second 15 min excursion to

a depth of 12 mH_2O is: $\xi = \xi\left(t = 45\,min; p_{O_2} = 160\,kPa\right) + \xi\left(t = 15\,min; p_{O_2} = 160\,kPa\right)$

$+ \xi\left(t = 45\,min; p_{O_2} = 160\,kPa\right) + \xi\left(t = 15\,min; p_{O_2} = 160\,kPa\right) \cong 0.094 \triangleq 9.4\%$ (Figure

7.9. Hazard at a level at the level of $\xi \cong 9\%$ usually is not accepted[28] for combat dives.

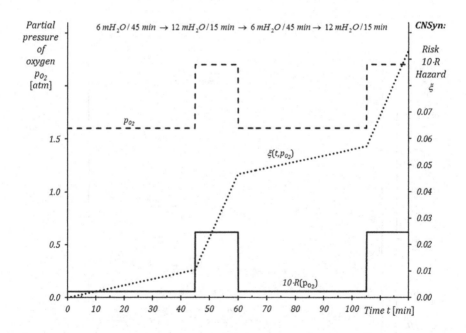

FIGURE 7.14 Dependence of risk R and the cumulative hazard function ξ of onset of *CNSyn* symptoms from the time t according to the semi-empirical algebraic mathematical model (4.26a) for a transit dive to 6 mH_2O with allowed 15 min excursion to 12 mH_2O and with unallowable second excursion to 12 mH_2O (Harabin A.L., Survanshi S.S., Homer L.D., 1995).

7.2.5 LONG TRANSIT DIVES

There were two lengthy transit dives without excursions performed at a depth of $6\ mH_2O$ (Figure A4.5. The time of dives was $\tau \in \{180; 199\}\ min$.

The average load expressed in pressure F exerted on the horizontal plate connected to tensiometer was $F \in \{2.12; 2.16\}\ kg$. According to Figure 7.3 it can be assumed that the average observed pressures F correspond to the average speeds on the distances covered by the swimmers expressed in knots $v \in \{0.8_1; 0.8_2\}\ kn$.

The heart rate during the exposure stabilized within $HR \in (92, 94)\ min^{-1}$.

The tests were carried out in water at a temperature $t \in \{15;17\}°C$ respectively.

After maintaining an approximately $(20, 25)\ min$ transit depth, the oxygen content in the circuit exceeded $C_{O_2} > 96\%_v$.

The maximum partial pressure during tests at a depth $H \in \{6; 6\}mH_2O$ with the corresponding time $\tau \in \{160; 200\}min$ amounted to $p_{O_2} \in \{157;157\}kPa$.

The final value of the cumulative hazard function ξ of onset of $CNSyn$ symptoms according to the algebraic semi-empirical mathematical model (4.26a) for transit with excursion dives to a depth of $6\ mH_2O$ is: $\xi = \xi\left(t = 200\ min, p_{O_2} = 160\ kPa\right) \cong 0.046$ $\triangleq 4.6\%$ and is accepted for combat dives (Figure 7.16).

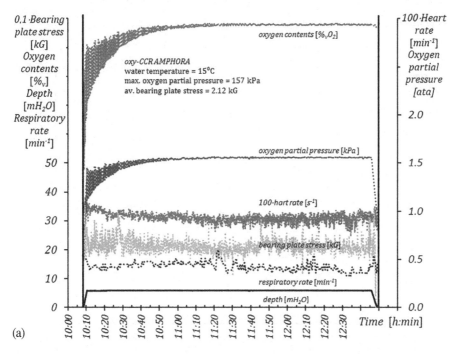

(a)

FIGURE 7.15 Results of oxygen lengthy transit dives without excursions performed at a depth of $6\ mH_2O$.

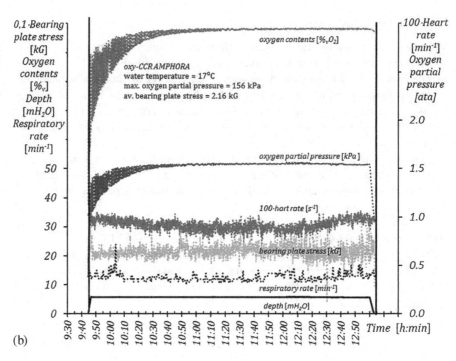

(b)

FIGURE 7.15   (Continued)

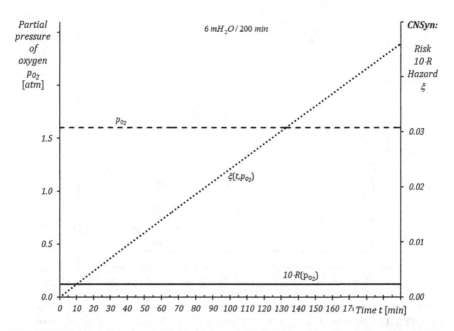

FIGURE 7.16   Dependence of risk R and the cumulative hazard function ξ of onset of *CNSyn* symptoms from time t according to the semi-empirical algebraic mathematical model (4.26a) for a transit dive to 6 mH_2O (Harabin A.L., Survanshi S.S., Homer L.D., 1995).

7.2.6 RISK OF PULMONARY TOXICITY

On completion of all the experimental oxygen dives, some parameters of the lung function were measured (Nunn J.F., 1993): VC – vital lung capacity [dm^3], FEV_1 – first minute forced expiratory volume [dm^3], FVC – forced vital lung volume [dm^3], PEF – peak expiratory flow [$dm^3 \cdot s^{-1}$], FEF 25–75 – concentrated expiratory flow counted between 25% and 75% of vital lung capacity [$dm^3 \cdot s^{-1}$], PIF – peak inspiratory flow [$dm^3 \cdot s^{-1}$], before, immediately after and after 1 h of oxygen exposure (Table 7.1).

During all the experimental oxygen dives a daily dose of pulmonary toxicity $UPTD$[29] was not been exceeded (Tables 7.1 and 3.7).

The statistical inference made using test t – $Student$ showed that the mean values computed from the differences between the initial values and those recorded after the dive are negligible.[30] Such a result does not authorize any further analysis of the data. However, it is worth noting that there is a tendency to compensate the minimum deviation of average values after a dive in relation to the initial values after 1 h of oxygen exposure, separately for individual divers and as a consequence also for the averaged data (Table 7.2).

An analysis of the measured lung function parameters is difficult not only because of statistically negligible deviations in the recorded magnitudes but also because of the strong covariance of some of these parameters (Appendix 3).

7.2.7 SUMMARY

As part of the experiments, allowed and not recommended oxygen diving procedures were tested using the diving apparatus in configuration oxy – $CCR\ AMPHORA$.

Due to the small available population of divers, studies were based only on random checking of assumed tactical situations. Oxygen excursions were taken not earlier than 45 min after a mission was carried out at the maximum transit depth of 6 mH_2O because an earlier excursion is unlikely from a tactical point of view. The mission times chosen were from 1 h and 15 min to 2 h. Three profiles were tested with an allowed oxygen excursion to a depth of 12 mH_2O and four profiles with an allowed oxygen excursion to a depth of 15 mH_2O. Two long allowed transit dives were performed at a depth of 6 mH_2O[31] (Table 7.3).

Four dives were performed with unallowable extension of excursion to 12 mH_2O and two dives with unallowable extension of excursion to 15 mH_2O. Double excursion scenarios with not recommended excursions to a depth of 12 mH_2O were also tested (Table 7.3).

Oxygen dives were carried out without prior purging of the breathing space of apparatus in the configuration of oxy – $CCR\ AMPHORA\ SCUBA$. Flushing was not performed since during the oxygen dives with a nitrox excursion, described later, the start of the diving process was preceded by intensive ventilation of the apparatus breathing space. Purging was performed to test the conditions of the worst scenario with onset of $CNSyn$. Whether the breathing space was flushed with oxygen or not a relatively rapid increase in concentration of oxygen[32] in the circuit was observed, so that hazard of $CNSyn$ symptoms was in both cases similar.

During all the 17 experimental dives, in no cases were symptoms of $CNSyn$ recorded. Neither were the symptoms of pulmonary toxicity were also not observed.

TABLE 7.1
The results of spirometry

UPTD	VC			FEV$_1$			FVC			PEF			FEF 25–75			PIF		
	before	after	after 1 h	before	after	after 1 h	before	after	after 1 h	Before	after	after 1 h	before	after	after 1 h	before	after	after 1 h
169	6.38	6.00	6.22	5.10	4.71	5.02	6.76	6.35	6.62	8.97	9.28	9.31	4.20	3.64	4.15	5.11	6.02	5.13
159	6.37	6.12	6.46	4.99	4.98	5.15	6.61	6.68	6.92	9.31	9.48	9.54	4.11	3.95	4.08	5.57	6.37	5.26
159	6.11	6.26	6.31	4.90	4.97	5.20	6.58	6.66	6.83	9.21	9.78	9.96	3.91	3.95	4.33	5.47	5.67	6.58
159	4.30	4.29	4.07	3.45	3.34	3.46	4.60	4.49	4.58	11.74	10.06	10.26	2.66	2.53	2.67	3.48	4.65	4.23
171	6.34	6.34	6.46	5.06	5.02	5.17	6.82	6.77	6.87	9.74	9.87	9.96	3.96	3.89	4.21	6.85	6.93	6.26
171	3.85	4.18	4.13	3.48	3.38	3.45	4.54	4.48	4.43	11.33	10.62	10.62	2.79	2.66	2.92	4.03	3.19	3.60
120	6.26	6.01	6.26	5.11	4.90	4.93	6.57	6.61	6.56	9.50	9.31	9.53	4.42	3.82	3.97	6.03	6.48	6.73
180	6.28	6.04	6.32	5.15	4.85	4.99	6.83	6.59	6.73	9.65	9.29	9.37	4.20	3.73	3.94	6.48	5.97	6.39
190	6.17	5.90	6.13	5.01	4.94	5.03	6.72	6.65	6.78	9.48	9.43	9.68	3.96	3.90	3.94	7.22	6.88	6.70
159	6.36	5.75	6.48	4.89	4.72	5.06	6.51	6.47	6.61	9.62	9.23	9.82	3.94	3.51	4.26	6.18	7.79	6.77
190	4.25	4.23	4.19	3.31	3.26	3.54	4.45	4.41	4.48	11.16	9.98	11.42	2.49	2.45	3.03	4.22	3.17	4.90
212	6.18	5.71	6.10	4.89	4.89	5.03	6.40	6.62	6.71	9.31	9.47	9.72	4.10	3.79	4.01	6.76	6.41	6.01
212	4.06	4.02	4.04	3.38	3.51	3.54	4.40	4.57	4.55	11.35	10.04	10.71	2.76	2.83	2.93	3.91	3.19	2.65
120	6.12	5.75	6.21	5.14	4.79	5.05	6.73	6.49	6.58	9.78	9.31	9.57	4.28	3.68	4.30	5.69	6.09	6.23
262	4.15	4.10	3.93	3.47	3.41	3.56	4.56	4.53	4.59	11.00	10.88	11.22	2.77	2.66	2.95	3.15	4.08	3.61
326	4.14	4.21	4.08	3.30	3.43	3.41	4.45	4.57	4.48	10.85	10.74	10.47	2.51	2.64	2.72	2.74	2.97	2.20
302	6.24	5.99	6.29	5.15	4.60	4.90	6.77	6.46	6.70	9.93	9.21	9.90	4.29	3.28	3.71	6.53	7.08	6.61

TABLE 7.2
The average results of 17 series of measurements of selected spirometry parameters for oxygen dives with uncertainty of the value of the average calculated from the distribution *t – Student* with the significance $\alpha_0 = 0.05$.

Parameter	Initial value	Value after exposure	Value 1 *h* after exposure
VC [dm^3]	5.3±0.6	5.2±0.5	5.3±0.6
FEV_1 [dm^3]	4.3±0.5	4.2±0.4	4.4±0.4
FVC [dm^3]	5.7±0.6	5.7±0.6	5.8±0.6
PEF [$dm^3 \cdot s^{-1}$]	10.3±0.5	9.8±0.3	10.2±0.4
FEF 25–75 [$dm^3 \cdot s^{-1}$]	3.5±0.4	3.2±0.3	3.5±0.4
PIF [$dm^3 \cdot s^{-1}$]	5.2±0.8	5.3±0.9	5.2±0.9

VC – vital lung capacity [dm^3]

FEV_1 – first minute forced expiratory volume [dm^3]

FVC – forced vital lung volume [dm^3]

PEF – peak expiratory flow [$dm^3 \cdot s^{-1}$]

FEF 25–75 – concentrated expiratory flow counted between 25% and 75% of vital lung capacity [$dm^3 \cdot s^{-1}$]

PIF – peak inspiratory flow [$dm^3 \cdot s^{-1}$]

TABLE 7.3
Summary of experimental oxygen dives performed with the use of the diving apparatus *oxy – CCR AMPHORA SCUBA*.

No	Figure	Diver	Dive profile Depth [mH_2O]/time [min]	Total diving time [min]
1.	7.5a	BRAVO	6/45 → 12/10 → 6/45	103
2.	7.6a	BRAVO	6/45 → 15/5 → 6/45	98
3.	7.6b	FOXTROT	6/45 → 15/5 → 6/45	98
4.	7.6c	BRAVO	6/45 → 15/5 → 6/45	98
5.	7.10a	BRAVO	6/45 → 15/10 → 6/45	104
6.	7.10b	FOXTROT	6/45 → 15/10 → 6/45	104
7.	7.5b	BRAVO	6/30 → 12/10 → 6/30	74
8.	7.9a	BRAVO	6/45 → 12/15 → 6/45	110
9.	7.9c	BRAVO	6/45 → 12/20 → 6/45	116
10.	7.6d	BRAVO	6/45 → 15/5 → 6/45	98
11.	7.9d	FOXTROT	6/45 → 12/20 → 6/45	156
12.	7.13a	BRAVO	6/45 → 12/15 → 6/45 → 12/15	125
13.	7.13b	FOXTROT	6/45 → 12/15 → 6/45 → 12/15	124
14.	7.15a	FOXTROT	6/160	163
15.	7.5c	BRAVO	6/30 → 12/10 → 6/30	75
16.	7.15b	FOXTROT	6/199	199
17.	7.9b	BRAVO	6/15 → 12/15 → 6/150	184

Dives conducted by the US Navy divers take place in different environmental conditions and with different equipment and with different diving gear. However, the large number of dives carried out by the US Navy and our own experience indicate that the US Navy technology of transit diving can be approved for use by the Polish Armed Forces' transit type diving technology. Of course, initially the dives should be done with the gradual complexity levels and under special supervision because of the small number of our own experimental dives and modest experience in technology of transit diving in the conditions of the Baltic Sea. Dives should be performed with similar limitations as experimental dives[33] until a minimum of 300 person-exposures are attained for each scenario of combat mission. After analyzing the results of these dives, it would be possible to design guidelines for planning diving operations with the use of oxygen as a breathing medium in a closed circuit apparatus.

7.3   EXPERIMENTAL OXYGEN/NITROX EXPOSURES

For the proposed technology of diving with oxygen as a breathing medium and the possibility of making an Nx excursion, one type of premix[34] Nx 0.43 was used. It was dosed at a level of $\dot{V} = 10.5 \, dm^3 \cdot min^{-1}$ (Table 1.4). This allowed running the oxygen experimental dives with an Nx excursion within the depth range of $(6,32) \, mH_2O$. For the apparatus $SCR \, AMPHORA \, SCUBA$ in configuration Nx, the maximum theoretical excursion time was set at $\tau \le 36 \, min$. When maintaining the pressure of emergency supply of Nx at $p = 5 \, MPa$, the time is reduced to $\tau \le 27 \, min$ (Table 1.5).

The process of diving started with preoxygenation[35] during which a diver's load rate on exercise simulator was within $(2,2.5) \, kG$ (Figure 7.3). After maintaining a 45 min transit depth, the supply was switched to Nx and the relief valve was opened.[36] Without flushing the breathing space, a quick dive to the desired depth was made with a speed[37] $(10,20) \, mH_2O \cdot min^{-1}$ and the load corresponding to about 3 kg, measured with a tensiometer, which simulated escape into the depth. The breathing medium was supplemented only when it was absolutely necessary, through a bypass[38] device. After reaching the excursion depth the load was reduced and maintained within (2, 2.5) kg. After the scheduled time elapsed, the supply of breathing medium was switched to oxygen and the relief valve[39] was closed. The breathing bag was emptied exhausting the breathing medium, in a controlled manner, to the water.[40] The procedure for emptying and purging the breathing space was modified during the research process. These differences were described in the analysis of individual blocks of experimental dives. Then, slowly at a speed ranged $(1,3) \, mH_2O \cdot min^{-1}$ and the load corresponding to approximately 2 kg, measured with a tensiometer, an ascent to the surface was made.[41] On the surface, the breathing space was ventilated with oxygen by conducting three purges. Ventilation involved quick closing of the oxygen cylinder in the apparatus in configuration $oxy - CCR \, SCUBA \, AMPHORA$, sucking a breathing medium through a mouthpiece and blowing it through the nose[42] out of the breathing medium recirculation circuit in the diving apparatus. Then the oxygen cylinder was opened, which allowed for filling up the breathing space with oxygen.[43]

These steps had to be repeated three times. After the purging, the divers descended to a depth of about $6 \, mH_2O$ at a rate ranging $(1,10) \, mH_2O \cdot min^{-1}$ and with a load of about 2 kg, measured with a tensiometer.[44] After completion of the dive at the depth of $6 \, mH_2O$ the oxygen cylinder had to be shut and the diver had to ascend to the surface at a rate ranging $(1,3)\,mH_2O\cdot min^{-1}$. When diving at a depth H greater than or equal to $H \geq 24\,mH_2O$ the decompression assumptions for[45] dives 18, 19, 21–24 were tested.[46] The transit which preceded the excursion was made at a depth of $3\,mH_2O$.[47] As previously noted, diving in large water bodies is accompanied by waves, which have significant impact on the diver swimming at a depth within the range 5–7 of wave height.[48] For this reason, remaining in such a water body above the depth of $3\,mH_2O$ causes serious problems. Hence, it was assumed that the worst conditions for employing preoxygenation would be at a depth of $3\,mH_2O$. The preoxygenation time was also chosen in relation to the adopted tactical situation. Assuming the maximum operating swim speed $\nu = 0.5\,kn$ and the tactical range $(700,800)\,m$, which requires possibly hidden activity due to the likelihood of detection, the minimum preoxygenation time was defined at the level of 45 min.

7.3.1 PRELIMINARY DIVING

Three preliminary dives were made to familiarize the experimental divers with the apparatus $oxy - CCR/mix - SCR\,AMPHORA\,SCUBA$ and to check all the system components necessary to carry out the experimental dives and collect measurements results.

The dives were conducted at a transit depth of $6\,mH_2O$ using the apparatus $oxy - CCR\,AMPHORA\,SCUBA$ and a single 10 min excursion to a depth $H = 15\,mH_2O$ using the apparatus $Nx - SCR\,AMPHORA\,SCUBA$. After the 10 min excursion, the diver returned to the transit depth of $6\,mH_2O$ and the mission was continued with $oxy - CCR\,AMPHORA\,SCUBA$ (Figure 7.17).

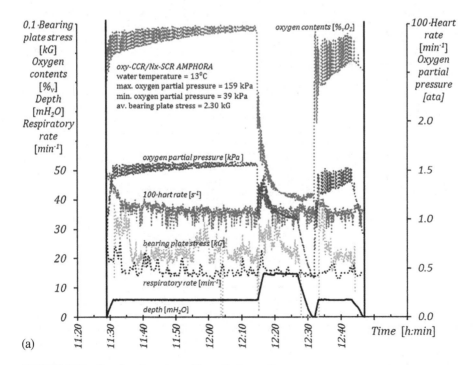

(a)

FIGURE 7.17   Results of $oxygen - Nx$ preliminary dives.

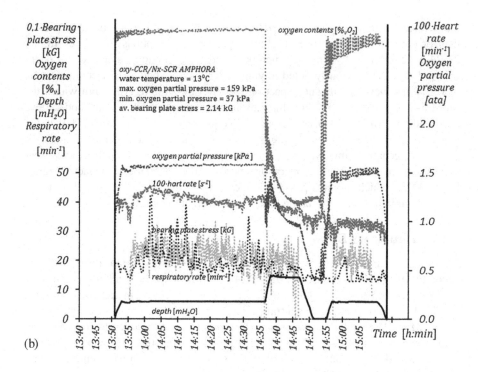

(b)

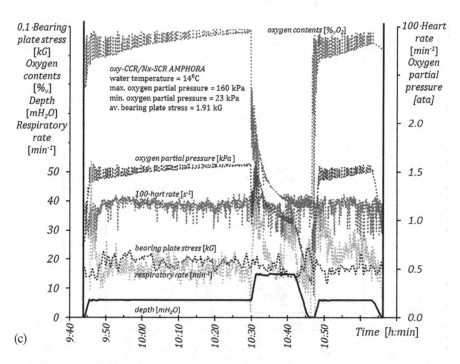

(c)

FIGURE 7.17   (Continued)

In the tested scenario, the excursion time was twice as long as that of the oxygen excursion time when the configuration $oxy - CCR$ $SCUBA$ $AMPHORA$ was employed.

An important element of the mission was to test the purge procedure when the breathing medium Nx was for oxygen. Prior to diving the purge procedure was executed, during which the breathing space was purged with oxygen three times on the surface. Prior to descent to the excursion depth, no purge of the breathing space was done, only the supply of breathing medium was changed to Nx and the relief valve was opened. The breathing medium was added in the circulation only if it was necessary. After the scheduled excursion time, the supply of breathing medium was switched to oxygen and the relief valve was closed. The breathing bag was emptied by exhausting the breathing medium, in a controlled manner, to the water.[49] Then a slow ascent was executed with normal breathing.

The average workload values for these dives, expressed in pressure F exerted on the horizontal plate connected to a tensiometer, were $F \in \{2.30; 2.14; 1.91\} kG$. According to Figure 7.3, the estimated average workload for the observed average pressure F corresponded to the average swim speeds at which the diver covered the distances expressed in knots $v \in \{0.8_5; 0.8_2; 0.7_7\} kn$.

The average heart rate during the exposure stabilized within $HR \in (110,116) min^{-1}$.

The dives were carried out in water at $t \in \{13; 13; 14\}°C$.

Theoretically, after the 45 min preoxygenation at a depth of $6 mH_2O$, no decompression is permissible after 100 min spent at a depth of $H = 15 mH_2O$ using Nx and the diving apparatus in the configuration $Nx - SCR$ $AMPHORA$ $SCUBA$, assuming that the saturation gradient is $\delta \cong 75\%$. If, after the excursion, the mission is continued at depths of up to $6 mH_2O$ using oxygen as the breathing medium in the configuration oxy of the diving apparatus $oxy - CCR$ $AMPHORA$ $SCUBA$ for 10 min, the excursion time may be extended to 176 min. However, these times are beyond the protective time of $Nx - SCR$ $AMPHORA$ $SCUBA$.

As expected, in none of these dives was any occurrence of intravascular free gas phase recorded at the end of a dive or one hour after the diving procedures were completed (Kłos R., 2010; Kłos R., 2011).

7.3.2 DIVES WITH EXCURSION TO 15 mH_2O

Seven dives were made in order to develop methods of changing the supply from Nx to oxygen.

The dives started with preoxygenation at a transit depth of $6 mH_2O$ using the diving apparatus in configuration $oxy - CCR$ $AMPHORA$ $SCUBA$ and oxygen as a breathing medium. Then a single 15 min excursion was made to a depth of $H = 15 mH_2O$ using the apparatus in configuration $Nx - SCR$ $AMPHORA$ $SCUBA$ and Nx as a breathing medium. After the excursion, the dive was continued for 20 min at a transit depth of $6 mH_2O$ using the apparatus in configuration oxy (Figure 7.18).

In the tested scenario, the excursion time was three times as long as that of when only the $oxy - CCR$ $AMPHORA$ $SCUBA$ was employed.

One of the aims of the mission was to test the purge procedure of the breathing space in the diving apparatus when changing the breathing medium from Nx to oxygen.

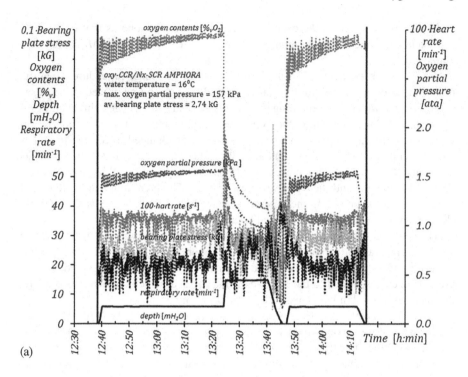

(a)

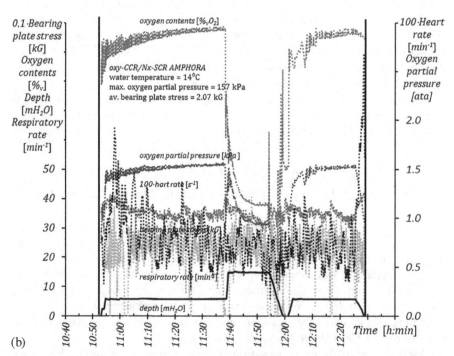

(b)

FIGURE 7.18   Results of *oxygen – Nx* dives with excursion to 15*mH₂O* depth.

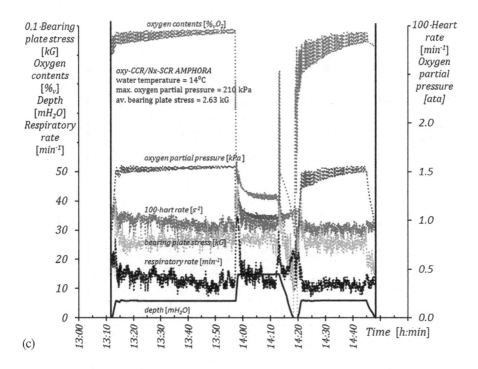

(c)

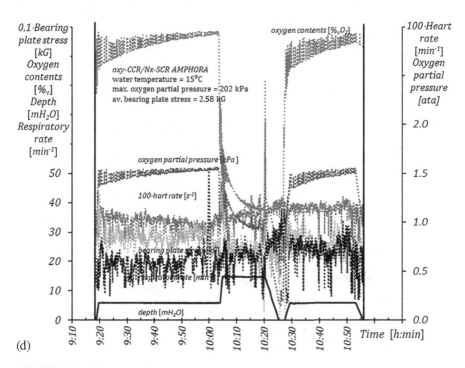

(d)

FIGURE 7.18 (Continued)

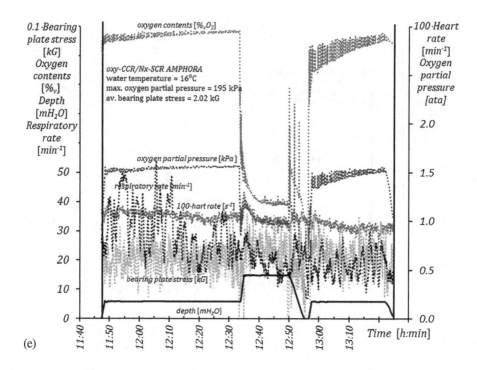

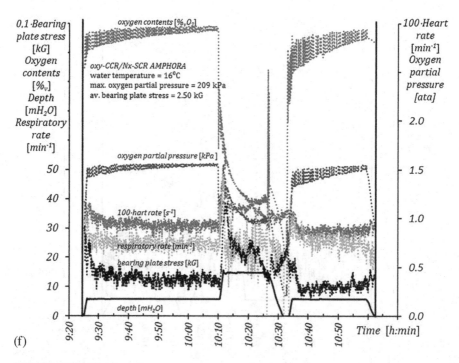

FIGURE 7.18 (Continued)

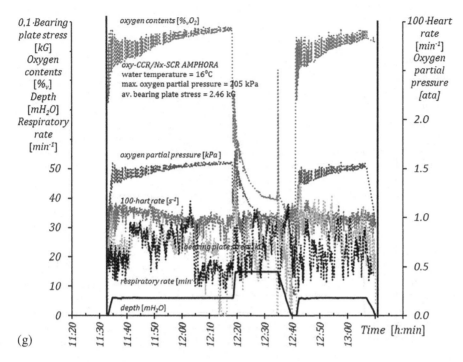

FIGURE 7.18 (Continued)

In the first step, the supply was switched to oxygen but the oxygen cylinders were not opened. Then the breathing bag was emptied and the oxygen cylinder was opened. Afterwards the ascent started.

The average loads for these dives, expressed in the pressure exerted on the horizontal plate connected to a tensiometer, were $F \in \{2.74; 2.07; 2.63; 2.58; 2.02; 2.50; 2.46\} kG$. According to Figure 7.3, the estimated average workload for the observed average pressure F corresponds to the average swim speeds at which the divers covered the distances, expressed in knots $v \in \{0.9_2; 0.8_0; 0.9_1; 0.9_0; 0.7_9; 0.8_8; 0.8_8\} kn$.

The divers' heart rates during the exposure stabilized in the range of $HR \in (96,106) min^{-1}$.

The dives were carried out in water at temperature $t \in \{13; 13; 14\}°C$.

Theoretically, no decompression is permissible for the following dive profile: 45 min at $6 mH_2O$ using $oxy - CCR\ AMPHORA\ SCUBA$ with a single excursion to a depth $H = 15 mH_2O$ using $Nx - SCR\ AMPHORA\ SCUBA$ and 20 min after surfacing, the continued oxygen dive to a depth of $6 mH_2O$, when the excursion time τ will is $\tau \le 234 min$, assuming that the saturation gradient is $\delta \cong 75\%$. However, this time is beyond the protective time of $Nx - SCR\ AMPHORA\ SCUBA$ (Table 1.5).

As expected, in none of the dives was any occurrence of intravascular free gas phase recorded at the end of dive or one hour after the diving procedures were completed (Kłos R., 2010; Kłos R., 2011).

One dive was also made with the at a transit depth of $6 mH_2O$ using $oxy - CCR$ $AMPHORA\ SCUBA$. Then there was a single 15 min excursion made to a depth of

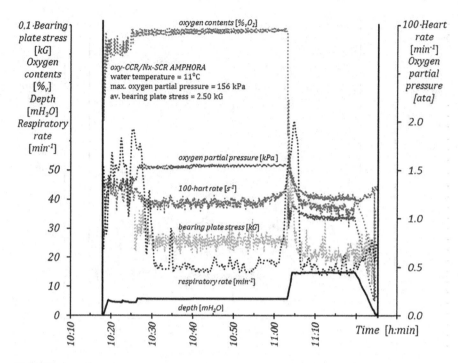

FIGURE 7.19   Result of *oxygen – Nx* dives with excursion to $15 mH_2O$ depth.

$H = 15 mH_2O$ using *Nx – SCR AMPHORA SCUBA*. Then decompression to the surface was conducted using *Nx – SCR AMPHORA SCUBA* (Figure 7.19).

The average workload for this dive, expressed in the pressure F, exerted on the horizontal plate connected to a tensiometer was $F \cong 2.50 kG$. According to Figure 7.3 the estimated average workload at pressure F corresponds to the average swim speed to cover the distance, expressed in knots $v \cong 0.8_8 kn$. The heart rate during the exposure stabilized at approximately $HR \cong 118 min^{-1}$. The dive was taken in water at temperature $t \cong 11°C$.

The aim of these dives was to confirm the safety of decompression assumptions. As expected, no occurrence of intravascular free gas phase was recorded at the end or one hour after the dive was completed (Kłos R., 2010; Kłos R., 2011).

7.3.3   Dives with Excursion to 24 mH_2O

Eleven dives were made to a depth of $24 mH_2O$. Two of them will not be analyzed here.[50] The first series of five experimental dives was made to test the diving apparatus purge procedures. The second series of six dives was made to test the safety of decompression.

The first series of five dives started at a transit depth $6 mH_2O$ using *oxy – CCR AMPHORA*. Then a single excursion was made to a depth of $H = 24 mH_2O$ using *Nx – SCR AMPHORA SCUBA*. After the excursion there was a return to a transit depth of $6 mH_2O$ and about $25 min$ of continuation of the mission using *oxy – CCR AMPHORA SCUBA* (Figure 7.20).

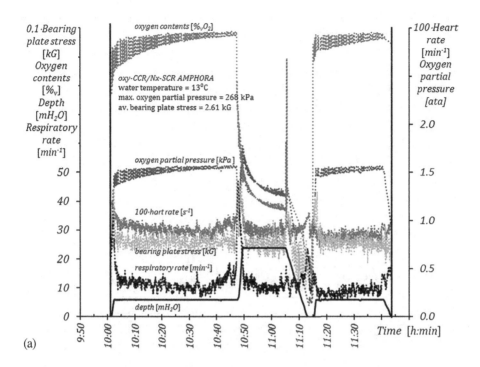

(a)

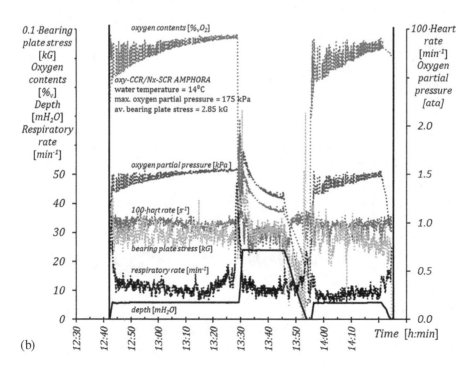

(b)

FIGURE 7.20   Results of *oxygen – Nx* dives with excursion to 24*mH₂O* depth.

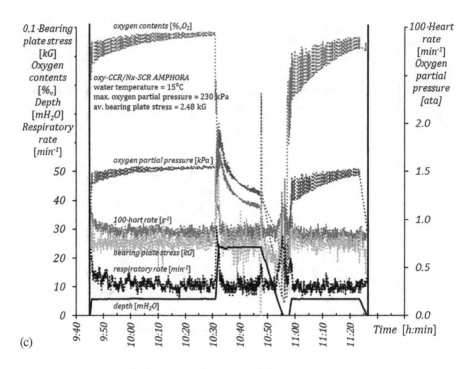

(c)

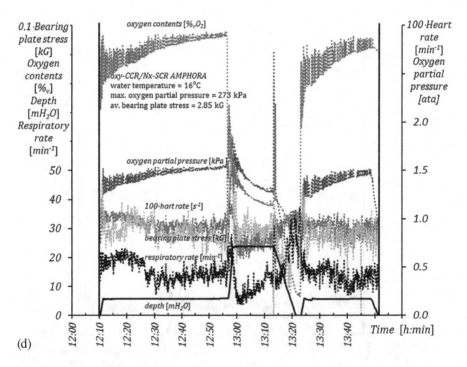

(d)

FIGURE 7.20 (Continued)

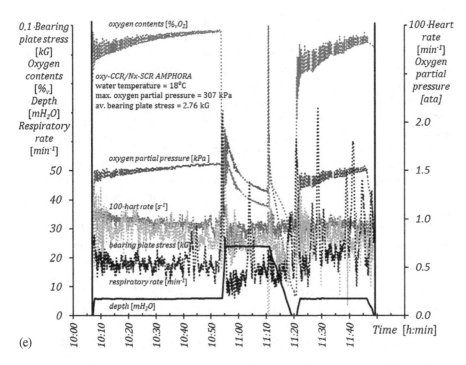

(e)

FIGURE 7.20 (Continued)

One of the aims of the mission was to test the purge procedures of the apparatus breathing space when changing the breathing medium from *Nx* to oxygen, modified in relation to the previous dives (Figures 7.20b–7.20e). The first step was to switch from *Nx* to oxygen, however, without opening the oxygen cylinder. Then the breathing bag was emptied, the cylinder with oxygen was opened, one breath[51] was taken and the relief valve was closed. Then the ascent started. For the dive referred to in Figure 7.20a, the purge procedure was the same as that for the excursions to the depth of $H = 15\,mH_2O$.

The average workload values for these dives, expressed in the pressure *F* exerted on the horizontal plate connected to a tensiometer, were $F \in \{2.61; 2.85; 2.48; 2.85; 2.76\}\,kG$. According to Figure 7.3, it can approximately be assumed that the average observed pressures *F* correspond to the average swim speeds at which the divers covered the distances, expressed in knots $v \in \{0.9_0; 0.9_4; 0.8_8; 0.9_4; 0.9_3\}\,kn$.

The divers' heart rates during the exposure stabilized within the range of $HR \in (88, 94)\,min^{-1}$.

The dives were conducted in water temperature $t \in \{13; 14; 15; 16; 18\}°C$ respectively.

Theoretically, no decompression is permissible for this dive profile: 45 *min* at $6\,mH_2O$ using $oxy - CCR\ AMPHORA\ SCUBA$ with a single 15 *min* excursion to a depth of $H = 24\,mH_2O$ using $Nx - SCR\ AMPHORA\ SCUBA$ and with a 20 *min* continuation of the oxygen dive to a depth of $6\,mH_2O$, when the excursion is less than $\tau < 83\,min$, assuming that the saturation gradient is $\delta \cong 75\%$. This time is beyond the protective time of $Nx - SCR\ AMPHORA\ SCUBA$ (Table 1.5).

As expected, in none of the dives was the intravascular free gas phase noticed after the dive or one hour after the end of the dive (Kłos R., 2010; Kłos R., 2011).

In the second round, six dives were conducted to verify the decompression assumptions, but only four of them were analyzed[52] (Figure 7.21).

The dives started with a 45 min exposure at a transit depth of 3 mH_2O using $oxy -$ $CCR\ AMPHORA\ SCUBA$. Then a single 30 min excursion was made to a depth of $H = 24\ mH_2O$ using $Nx - SCR\ AMPHORA\ SCUBA$. After the excursion, decompression to the surface was executed.

One of the aims of the mission was to test a modified purge procedure when changing the breathing medium Nx for oxygen. Prior to the purge, the supply breathing medium was changed to oxygen, but without opening the oxygen cylinder. Then the breathing bag was emptied, oxygen cylinder opened, one breath[53] was taken and the relief valve was closed. Then the ascent started.

The average workload values for these dives, expressed in pressure F exerted on the horizontal plate connected to a tensiometer, were $F \in \{1.81; 2.04; 2.44; 1.95\}\ kG$. According to Figure 7.3, it can approximately be assumed that the average observed pressures F correspond to the average swim speeds at which the divers covered the distances, expressed in knots $v \in \{0.7_5; 0.8_0; 0.8_7; 0.7_8\}\ kn$.

The divers' heart rates during the exposure stabilized within the range of $HR \in (101; 113)\ min^{-1}$.

The dives were conducted in water temperature $t \in \{18; 18; 19; 18\}\ °C$ respectively.

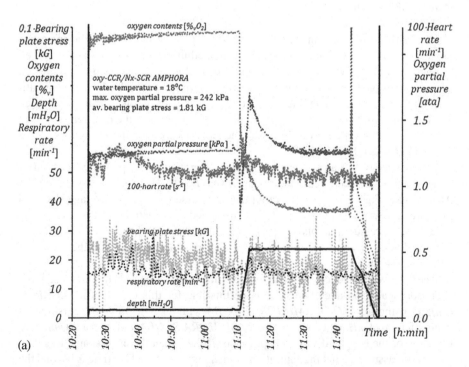

FIGURE 7.21 Results of $oxygen - Nx$ dives with excursion to $24mH_2O$ depth.

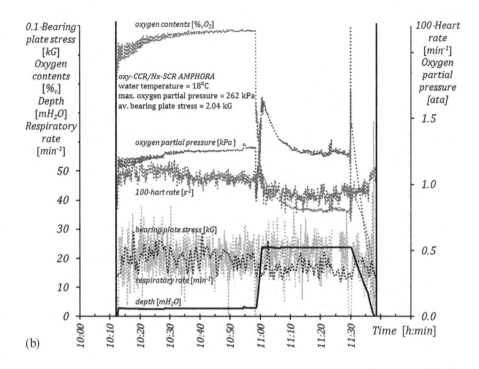

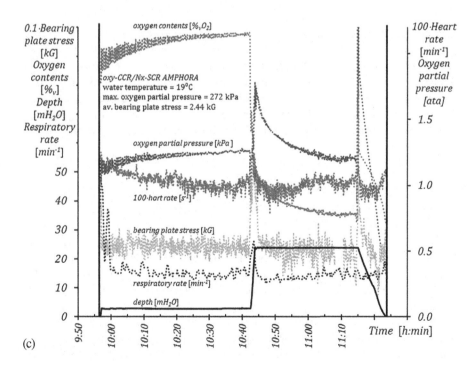

FIGURE 7.21   (Continued)

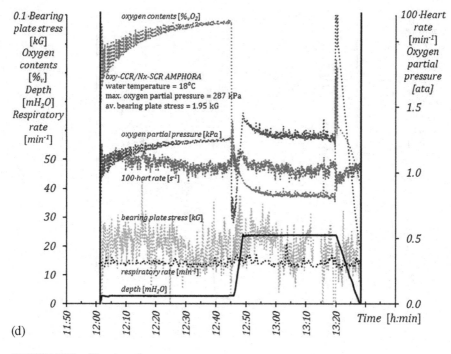

(d)

FIGURE 7.21 (Continued)

Theoretically, no decompression is permissible for this dive profile: 45 *min* at $3\,mH_2O$ using *oxy* – *CCR AMPHORA SCUBA* with a single excursion to a depth $H = 24\,mH_2O$ using – *CCR AMPHORA SCUBA*, when the excursion time is $\tau \le 25\,min$ and assuming that the saturation gradient is $\delta \cong 75\%$. The excursion time $\tau = 30\,min$ generates 80% saturation gradient for tissue number 2 and is acceptable. As expected, in none of the dives was occurrence of the intravascular free gas phase noticed after the dive one hour after the diving procedures were completed (Kłos R., 2010; Kłos R., 2011).

7.3.4 DIVES WITH EXCURSION TO 32 mH_2O

Two dives were made to a depth of $32\,mH_2O$ to verify the decompression assumptions (Figure 7.22).

The dives started at a transit depth of $3\,mH_2O$ using *oxy* – *CCR AMPHORA SCUBA*. After 45 *min* during the first dive and 20 *min* during the second one, a single excursion was made to a depth of $H = 32\,mH_2O$ using *Nx* – *SCR SCUBA AMPHORA*. After the excursion was completed, decompression to the surface was executed.

One of the aims of the mission was to test the purge procedures when the breathing medium was changed from *Nx* to oxygen. First, the supply was switched to oxygen, but without opening the oxygen cylinder. Then the breathing bag was emptied by pressing the apparatus to the diver's body. After emptying the bag, the oxygen

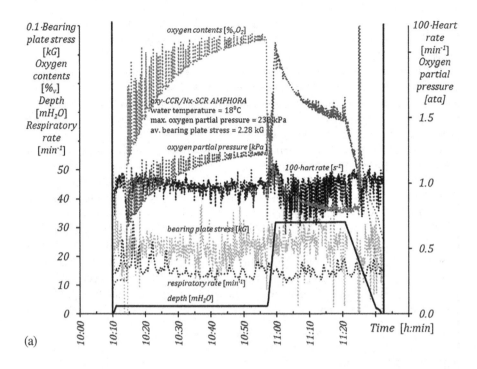

(a)

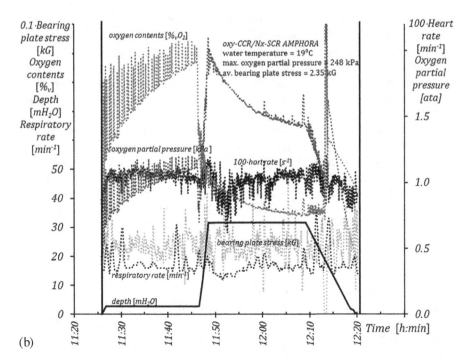

(b)

FIGURE 7.22   Results of *oxygen – Nx* dives with excursion to *32mH₂O* depth.

cylinder was opened and one breath was taken.[54] Then the breathing bag was emptied again by pressing it to the diver's body and the relief valve was closed. After that the ascent began.

The average workload values for these dives, expressed in pressure F exerted on the horizontal plate connected to a tensiometer, were $F \in \{2.28; 2.35\} kG$. According to Figure 7.3, it can approximately be assumed that the average observed pressures F correspond to the average swim speeds at which the distances were covered, expressed in knots $v \in \{0.8_4; 0.8_6\} kn$.

The divers' heart rates during the exposure stabilized in the range of $HR \in (97, 102) min^{-1}$.

The dives were conducted in water temperature $t \in \{18; 19\}°C$ respectively.

Theoretically, no decompression is permissible for this dive profile: 45 *min* at $3 mH_2O$ using $oxy - CCR$ AMPHORA SCUBA with a single excursion to a depth of $H = 32 mH_2O$ using $Nx-$ CCR AMPHORA SCUBA when the excursion time τ is $\leq 13 min$, assuming that the saturation gradient is $\delta \cong 75\%$. An excursion time of approximately 20 *min* generates 95% saturation gradient for tissue number 2.

For trained divers carrying out combat missions, this time is also acceptable.

As expected, in none of dives was occurrence of intravascular free gas phase noticed after the dive or one hour after the diving procedures were completed (Kłos R., 2010; Kłos R., 2011).

7.4 CONCLUSION

Three preliminary dives were made to familiarize experimental divers with the diving apparatus $oxy - CCR/mix$ SCR AMPHORA SCUBA and to check all the systems employed for diving and collecting results. The dives were made at a transit depth of $6 mH_2O$ using $oxy - CCR$ AMPHORA SCUBA with a single 10 *min* excursion to a depth of $H = 15 mH_2O$. After the excursion, there was return to the transit depth of $6 mH_2O$ and a 10 *min* continuation of the mission using $oxy - CCR$ AMPHORA SCUBA.

Seven dives were made to a depth of $H = 15 mH_2O$ to develop methods of transition from oxygen to Nx. The dives started with preoxygenation at a transit depth of $6 mH_2O$ using the diving apparatus CCR AMPHORA SCUBA in configuration oxy and oxygen as a breathing factor. Then a single 15 *min* excursion was made to a depth of $H = 15 mH_2O$ using the diving apparatus SCR AMPHORA SCUBA in configuration Nx and nitrox as a breathing medium. After the excursion, the dive continued for 20 *min* at a transit depth of $6 mH_2O$ using the diving apparatus $oxy -$ CCR AMPHORA SCUBA in configuration.

Eleven dives were made to a depth of $24 mH_2O$. Two of them were not analyzed here: numbers 17 and 20 from Table 7.4.

The first series of five dives was run to test the diving apparatus purging procedures. The second series of six dives was to test the safety of decompression. The first series of five dives started at a transit depth $6 mH_2O$ using $oxy - CCR$ AMPHORA SCUBA. Then a single 15 *min* excursion to a depth of $H = 24 mH_2O$ was made using $Nx - SCR$ AMPHORA SCUBA. After the excursion, there was a return to the transit depth of $6 mH_2O$ and an approximately 20 *min* continuation of the mission

TABLE 7.4
List of experimental oxygen/Nx dives using *oxy – CCR/Nx SCR AMPHORA SCUBA*.

No.	Fig. no.	Diver	Dive profile Direction/depth/time				Total dive time
				[1]/[mH_2O]/[min]			[min]
1.	7.17a	FOXTROT	O_2	↓6/45	—	↓6/10↑	78
1.			Nx	—	↓15/10↑	—	
2.	7.17b	ALFA	O_2	↓6/45	—	↓6/15↑	83
1.			Nx	—	↓15/10↑	—	
3.	7.17c	BRAVO	O_2	↓6/45	—	↓6/15↑	83
1.			Nx	—	↓15/10↑	—	
4.	7.19	FOXTROT	O_2	↓6/45	—	—	72
1.			Nx	—	↓15/15↑	—	
5.	7.18a	BRAVO	O_2	↓6/45	—	↓6/25↑	87
1.			Nx	—	↓15/15↑	—	
6.	7.18b	HOTEL	O_2	↓6/45	—	↓6/25↑	96
1.			Nx	—	↓15/15↑	—	
7.	7.18c	FOXTROT	O_2	↓6/45	—	↓6/25↑	96
1.			Nx	—	↓15/15↑	—	
8.	7.18d	BRAVO	O_2	↓6/45	—	↓6/25↑	98
1.			Nx	—	↓15/15↑	—	
9.	7.18e	HOTEL	O_2	↓6/45	—	↓6/25↑	98
1.			Nx	—	↓15/15↑	—	
10.	7.18f	FOXTROT	O_2	↓6/45	—	↓6/25↑	97
1.			Nx	—	↓15/15↑	—	
11.	7.18g	BRAVO	O_2	↓6/45	—	↓6/25↑	97
1.			Nx	—	↓15/15↑	—	
12.	7.20a	FOXTROT	O_2	↓6/45	—	↓6/25↑	103
1.			Nx	—	↓24/15↑	—	
13.	7.20b	BRAVO	O_2	↓6/45	—	↓6/25↑	102
1.			Nx	—	↓24/15↑	—	
14.	7.20c	FOXTROT	O_2	↓6/45	—	↓6/25↑	103
1.			Nx	—	↓24/15↑	—	
15.	7.20d	BRAVO	O_2	↓6/45	—	↓6/25↑	102
1.			Nx	—	↓24/15↑	—	

TABLE 7.4 *(Continued)*

List of experimental oxygen/Nx dives using *oxy – CCR/Nx SCR AMPHORA SCUBA*.

No.	Fig. no.	Diver	Dive profile Direction/depth/time				Total dive time
				[1]/[mH_2O]/[min]			[min]
16.	7.20e	BRAVO	O_2	↓6/45	—	↓6/25↑	102
1.			Nx	—	↓24/15↑	—	
17.	—	HOTEL	O_2	↓3/45	—	—	78
1.			Nx	—	↓24/28↑	—	
18.	7.21a	ALFA	O_2	↓3/45	—	—	89
1.			Nx	—	↓24/30↑	—	
19.	7.21b	HOTEL	O_2	↓3/45	—	—	86
1.			Nx	—	↓24/30↑	—	
20.	—	ALFA	O_2	↓3/45	—	—	87
1.			Nx	—	↓24/30↑	—	
21.	7.21c	FOXTROT	O_2	↓3/45	—	—	87
1.			Nx	—	↓24/30↑	—	
22.	7.21d	ALFA	O_2	↓3/45	—	—	87
1.			Nx	—	↓24/30↑	—	
23.	7.22a	BRAVO	O_2	↓3/45	—	—	82
1.			Nx	—	↓32/20↑	—	
24.	7.22b	BRAVO	O_2	↓3/20	—	—	55
1.			Nx	—	↓32/20↑	—	

using *oxy – CCR AMPHORA SCUBA*. In the second series six dives were made, but only four of them, which were used to verify the decompression assumptions, were analyzed. The dives started with a 45 *min* exposure to a transit depth of $3\,mH_2O$ using *oxy – CCR AMPHORA SCUBA*. Then a single 30 *min* excursion to a depth of $H = 24\,mH_2O$ was made using *Nx– SCR AMPHORA SCUBA*. After the excursion was completed, the decompression procedure to the surface was executed.

Two dives were made to a depth of $32\,mH_2O$ to verify the decompression assumptions. The dives started at a transit depth of $3\,mH_2O$ using *oxy – CCR AMPHORA SCUBA*. Then a single 20 *min* excursion was made to a depth of $H = 32\,mH_2O$ using *Nx – SCR AMPHORA SCUBA*. After the excursion, the decompression to the surface was executed.

In total, 24 experimental dives were made. They are a basis for a proposal extending the oxygen diving limits (procedure) by nitrox excursions.

NOTES

1 These functions can be individually switched off.
2 Built-in breathing system.
3 Before the return valve of mouthpiece – Figure 1.3.
4 The carbon dioxide content and oxidation compounds were also monitored.
5 Indirect measurement of effort – as shown later.
6 Mostly, used backup time was $2s$.
7 It may also be any portable computer connected to sockets in the Ethernet network of the DGKN-120 complex.
8 For example during the experiments with Nx - $SCRAMPHORASCUBA$ the test diver breathes oxygen or nitrox with oxygen content of $C_{O_2} \geq 30\%_v O_2 / N_2$, while the standby diver breathes air with oxygen content of $C_{O_2} \cong 21\%_v O_2 / N_2$.
9 Each chamber in DGKN-120 has two hatches, and space between them can be independently evacuated to the atmospheric pressure or connected with the neighboring chamber to equalize the pressures.
10 Transfer to the living chamber guarantees that the moment accumulation of intravascular free gas phase is detected, the medical recompression can be quickly undertaken.
11 Preferred time is less than 5 min.
12 Unless the doctor supervising the procedure decided otherwise, but minimum observation time could not be less than 1.5 h.
13 Especially neurological.
14 Because of the possibility of falsifying the results due to excessive overloading or adaptation to the conditions of the experimental dives.
15 The diver could also participate in the exposures as a safety diver up to three times during the six-day working week.
16 The formula has been derived previously with the method of dimensional analysis (Kłos R., 2011).
17 Coefficient has been taken from the literature (Troskalański A.T., 1954).
18 Due to the approximate nature of calculations, the density of distilled water was adopted here $\rho = 1000 \ kg \cdot m^{-3}$.
19 Flushing body with oxygen – in normal circumstances nitrogen dissolved in tissues is in equilibrium with the atmospheric air, but during the initial phase of the oxygen dive it is flushed from the tissues, causing delay in the dissolution of nitrogen in the tissues during the phase of the dive. Taking this delay into account allows for a slight shortening of the decompression process without increased risk of DCS.
20 Depending on how fast of a compression the diver withstood.
21 Simulation of controlled escape.
22 It was necessary to be careful and prevent gas emissions to water – to keep the secrecy of action.
23 It was necessary to prevent emissions of bubbles in large groups.
24 Now the aim is that the diver breathes nitrox in which concentration of oxygen during the dive should not exceed the limit of $85\%_v O_2 / N_2$ (Harabin A.L., Survanshi S.S., Homer L.D., 1994; Walters K.C., Gould M.T., Bachrach E.A., Butler F. K., 2000).
25 For dives Figure 7.5b, the data is missing due to a malfunction in the system recording the heart and breathing rate.
26 For dives in Figure A4.3a, the data is missing due to a malfunction in the system recording the heart and breathing rate.
27 The hazard at a level below $\xi \leq 5\%$ is accepted.

28 the threat at a level of below $\xi \leq 5\%$ is accepted.
29 because the interval between the dives exceeded a total compensation occurred – Table 3.7.
30 Inference was preceded by *F-Snedecora* test, which showed that the observed variances do not differ much from each other at a significance level $\alpha_0 = 0.05$, allowed doing a test *t-Student*.
31 The only limit was the time of protective action of the Co_2 absorber – Chapter 1.
32 An increase in oxygen content was caused by ventilation of the breathing space performed to collect a sample from the breathing atmosphere of the apparatus.
33 The best solution would be to conduct our own research in order to develop the optimal rules for oxygen dives together with the guidelines for using them for tactical actions.
34 $Nx \cdot (43.0 \pm 0.5)\%_v\, O_2\,/\,N_2$; dosing $\dot{V} = (10.5 \pm 1.0)\,dm^3$.
35 During the phase of diving with diving apparatus in configuration *oxy-CCR SCUBA AMPHORA* and oxygen as a breathing medium.
36 Configurations of diving apparatus *Nx-SCR AMPHORA SCUBA*.
37 Depending on how fast of a compression the diver could withstand.
38 It was necessary not to release gas into water to keep the action clandestine.
39 configuration *oxy-CCR AMPHORA SCUBA*.
40 The intention was not to cause emissions of large bubbles in dense groups.
41 To assess the situation and to rinse apparatus on the surface in order not to make noise in the water.
42 Or with mouth after closing the mouthpiece.
43 At this time the apparatus was not used for breathing.
44 Return to the continuation of the mission at the transit depth.
45 After excursion decompression up to the surface was conducted.
46 See later in this chapter – Table 7.4.
47 Minimum preoxygenation.
48 If the wave height is X, then the wave has a significant effect on the diver who is located at depths up to five/seven times wave height 5·X–7·X.
49 The intention was not to cause emissions of large bubbles in dense groups.
50 See numbers 17 and 20 in Table 7.4.
51 Inspiration and expiration.
52 The first was rejected because the breathing medium had been exhausted, the second was rejected because of a failure of the data recording; however, in neither of these cases was the intravascular free gas phase noticed.
53 Inspiration and expiration.
54 Inspiration and expiration.

REFERENCES

Harabin AL, Survanshi SS & Homer LD. 1994. *A Model for Predicting Central Nervous System Toxicity from Hyperbaric Oxygen Exposure in Man: Effects of Immersion, Exercise, and Old and New Data.* Bethesda: Naval Medical Research Institute. NMRI 94-0003; AD-A278 348.
Harabin AL, Survanshi SS & Homer LD. 1995. A model for predicting central nervous system toxicity from hyperbaric oxygen exposure in humans. *Toxicol. Appl. Pharmacol.*, 132, 19–26.
Kłos R. 2010. Intravascular bubble detection. *Pol. Hyperb. Res.*, 32, 15–30.

Kłos R. 2011. *Możliwości doboru dekompresji dla aparatu nurkowego typu CRABE – założenia do nurkowań standardowych i eksperymentalnych.* Gdynia: Polskie Towarzystwo Medycyny i Techniki Hiperbarycznej. ISBN 978-83-924989-4-0.

Kłos R. 2020. Ultrasonic detection of the intravascular free gas phase in research on diving. *Pol. Marit. Res.*, 106, 176–186. ISSN 1233-2585, e-ISSN 2083-7429, DOI: 10.2478/pomr-2020-0039.

Nunn JF. 1993. *Nunn's Applied Respiratory Physiology.* Jordan Hill: Butterworth-Heinemann Ltd. ISBN 0 7506 1336 X.

Troskalański AT. 1954. *Hydromechanika techniczna-Hydraulika.* Warszawa: Państwowe Wydawnictwo Techniczne.

US Navy Diving Manual. 2016. *The Direction of Commander: Naval Sea Systems Command* (revision 7). Washington. SS521-AG-PRO-010 0910-LP-115-1921.

Walters KC, Gould MT, Bachrach EA & Butler FK. 2000. Screening for oxygen sensitivity in US Navy combat swimmers. *Undersea Hyper. Med.*, 27, 21–26.

8 Remarks

The research program focused on two areas. The first was to develop a method of oxygen[1] dives using transition with a single excursion limit and to check the possibility of taking one more excursion, completing the dive.

Another scenario focused on testing the possibility of taking a nitrox (*Nx*) excursion after the initial phase of transition on oxygen when the diver was using the apparatus *oxy – CCR/Nx – SCR AMPHORA SCUBA* in *oxy mode*. In the excursion phase the supply of oxygen was changed to *Nx*, employing the semiclosed circuit breathing apparatus *Nx – SCC AMPHORA SCUBA*.

8.1 SCOPE OF WORK

Chapter 1 describes some of the important properties of the object investigated in the research that affect the safety of hyperbaric exposures. The object studied was a diving apparatus *oxy – CCR/Nx – SCR AMPHORA SCUBA*.

Selecting adequate mathematical models of respiratory ventilation of the diving apparatus is essential prior to the experimental work with human subjects. A model of ventilation of a semiclosed circuit diving apparatus with *Nx* as a breathing medium is a direct basis for planning a safe decompression process. A mathematical model of ventilation can also be used to develop procedures for effective purging of the respiratory loop of an apparatus, which is the basis for its safe operation and applicability of the emergency procedures used to stabilize its work.

The duration of the apparatus *oxy – CCR/Nx – SCR AMPHORA* is limited by the capacity of the carbon dioxide CO_2 absorber preparations contained in the canister to regenerate the breathing medium. Thus, the maximum time of a single mission τ is set at $\tau \leq 150\,min$. In exceptional cases the time may not exceed $\tau < 200\,min$.

In Chapter 2, we propose use of the oxygen window theory to model the mechanisms of central oxygen toxicity – *central nervous syndrome (CNSyn)*. Increased tension of oxygen physically dissolved in the blood was assumed as a driving force of the reactions leading to the formation of potentially toxic metabolites. The theory of oxygen window has traditionally been used for planning and safety assessment of decompression procedures as well as for planning hyperbaric treatments. This theory also provides a bridge between the phenomena occurring in the decompression after both short and saturation dives. The concept of the *extended oxygen window* was employed by Professor T. Doboszyński to develop continuous decompression tables applicable to *nitrox (Nx)* and *trimix (Tx)* saturation dives, which turned out to be an important Polish contribution to diving physiology theories.

The use of the concept of the *oxygen window* to describe the phenomenon of oxygen toxicity has not been widely published, but it was concentrated on the decompression process (Behnke A.R., 1967; Brubakk A.O., Neuman T.S., 2003). Therefore, the theories presented in this chapter are not reflected in other reports, so they should be treated with caution. They are an attempt in theoretical interpretation of the occurrence of *CNSyn*. The chapter includes a suggestion that the exceeded threshold value of the oxygen partial pressure in the breathing atmosphere, which causes complete saturation of hemoglobin with oxygen, reduces the time this atmosphere can be used safely for breathing because of *CNSyn*.

Chapter 3 briefly discusses the toxic effect of oxygen, divided into central nervous, lung and somatic toxicity together with the publications referred to. It points to some toxicity symptoms that accompany the use of oxygen as a breathing medium in diving.

In Chapter 4 we present a theoretical introduction to survival analysis. Proposed by the US Navy, the model of predicting the risk of occurrence of *CNSyn*, built on the basis of this theory, seems to be sufficiently precise. In the proposed model, the *central nervous system* (*CNS*) toxicity dose accumulates all the time. Hence, the changes in *CNS* toxicity dose are independent of the sequence of phases of a dive. This theory was used to plan the experimental dives, to make changes to the technology of diving on oxygen devised by the US Navy and to propose the technology of oxygen diving with allowable nitrox excursions to a depth of $24\,mH_2O$.

It seems strange that the technology of oxygen diving adopted by the US Navy does not take into account the results of the *CNSyn* risk analyses done and published by the authors of this technology.

Using the methods described in Chapter 4, we propose changes in the rules applicable to dives using *oxy – CCR/Nx – SCR AMPHORA SCUBA*. It appears that the proposed theory can be generalized to any exposure, depending on the acceptable level of *CNSyn* symptoms. Chapter 5 introduces tactical considerations and contains assumptions to the experiments designed to verify technology of diving using oxygen as a breathing medium.

Chapter 6 describes the possibilities of *Nx* excursions with no decompression as an extension of the technology of oxygen dives conducted within the transition with excursion limits.

Chapter 7 describes the results of experimental dives. The previously developed guidelines and experimental dives conducted within these guidelines paved the way to developing a technology of oxygen diving with allowable oxygen/*Nx* excursions. This study checked the method used to purge the breathing loop of the apparatus and verified the safety of the decompression procedures. The proposed technology of diving covers the most likely tactical scenarios (see Appendix 1).

8.2 TACTICAL CONSIDERATIONS

Combat diving with the use of oxygen is performed with the goal of silently transferring a special ops group/section. Normally dives are carried out at the small depth of $6\,mH_2O$. The choice of such a depth is forced not only by the toxic effects of oxygen, but also by tactical reckoning.

Passive technical means used for detecting combat divers are often placed at the bottom,[2] and their performance drastically decreases with distance. The risk of being detected by these means is reduced by keeping close to the surface. Some active diver detection means, such as sonar systems, demonstrate reduced detection effectiveness at the surface.

To develop the guidelines, we adopted an operating radius of 800 m as the smallest at the approach of the unreconnoitered shore because of possible protective/surveillance technical infrastructure installed on shore. Hence, it was assumed that the excursion phase after an oxygen transit at a depth of $6\,mH_2O$ could be expected no earlier than after a 45 min interval, assuming that the maximum swim speed does not exceed 0.5 kn.

The decompression assumptions were tested during escape dives into depth of water, using Nx as a breathing medium. Therefore, the transit part was conducted at a depth of $3\,mH_2O$. Diving in large bodies of water is characterized by waves, which significantly affect a diver swimming at a depth within 5–7 of the wave height.[3] For this reason, to maintain a depth shallower than $3\,mH_2O$ in such condition causes serious problems. Hence, it was assumed that diving at a depth of $3\,mH_2O$ would be characterized by the worst conditions for preoxygenation. The preoxygenation time was also chosen in conjunction with the adopted tactical situation. Given the risk for a diver to be detected, for the same swim speed magnitudes as mentioned earlier and the same tactical range, the preoxygenation minimum time will be approximately 45 min.

8.3   EXPERIMENTAL DIVING

As part of this study, the procedures of oxygen diving were tested. Seventeen experimental oxygen dives were conducted using the apparatus $oxy - CCR/Nx - SCR$ $AMPHORA\ SCUBA$ in the mode described in Chapter 5 as either allowed or unallowable. Twenty-four experimental oxygen dives with Nx excursions to a maximum depth of $32\,mH_2O$ were conducted using the configuration $oxy - CCR/Nx - SCR$ $AMPHORA\ SCUBA$.

8.4   CONCLUSIONS

Within the frame of this project, the oxygen diving procedures with Nx excursions to a maximum depth of $32\,mH_2O$ were tested. Due to the small population of divers available for the study, we performed only random verification of the assumed tactical situations. The theoretical studies and experimental dives, however, provided grounds for proposing extension of the limits for oxygen dives with nitrox excursions, thus increasing the flexibility of the combat operations (see Appendix 1).

The proposed technology of oxy/Nx dives using the apparatus $oxy - CCR/Nx -$ $SCR\ AMPHORA\ SCUBA$ does not exhaust, important from a tactical point of view, other scenarios of combat missions. It should be further supplemented by other essential elements. For example, important aspects of a combat situation may include safe transport by helicopter after diving missions, the impact of physical activity in a combat mission effort on safe withdrawal from the mission area by swimming underwater, the impact of underwater of transport on the level of physical activity in combat and so on. Learning the impact, made by adaptation training, on changes in individual resistance to decompression or exposure to high oxygen partial pressures is not without significance.

NOTES

1   Using the diving apparatus *oxy – CCR AMPHORA SCUBA.*
2   Like magnetic sensors.
3   If the wave height is X, the wave has a significant impact on the diver who is at a depth
    of up to five/seven times the wave height 5·X–7·X.

REFERENCES

Behnke AR. 1967. The isobaric (oxygen window) principle of decompression. In *The New Thrust Seaward. Transactions of the Third Annual Conference of the Marine Technology Society Conference, San Diego.* 5–7 June. Washington, DC: Marine Technology Society.

Brubakk AO & Neuman TS. 2003. *Bennett and Elliott's Physiology and Medicine of Diving.* London: Saunders. ISBN 0-7020-2571-2.

Appendix 1
Guidelines for Oxygen Dives with the Possibility of an Oxygen/Nitrox Excursion

i. Dives can be conducted with the self-contained underwater breathing apparatus $Nx - SCR/oxy - CCR$ *AMPHORA SCUBA* and other auxiliary diving gear designed for use with the aforementioned apparatus operated by the armed forces.

ii. Dives should be conducted in pairs and in compliance with applicable diving regulations. A good practice is to develop a diving manual for each operation where the $Nx - SCR/oxy - CCR$ *AMPHORA SCUBA* is used, consulting this document with the competent authority responsible for the safety of diving in the units subordinate to the Ministry of Defense (*MoD*).

iii. It is possible to use the oxygen diving procedure without nitrox excursions for diving apparatuses type $oxy - CCR$ *OxyNG SCUBA*, $oxy - CCR$ *FROG SCUBA*, $oxy - CCR$ *CODE SCUBA* and with auxiliary diving equipment designed for use with the aforementioned apparatuses operated by the armed forces. However, the maximum time of protective effect of these systems should be taken into account.

iv. Selection of dive profiles should be based on the principles described later, which, after authorization by the appropriate command, should be treated as a supplementary manual for planning training and combat missions.

v. The ability to regenerate respiratory medium with a carbon dioxide (CO_2) absorbent canister is a limiting factor on the duration of protective effect of the apparatus during oxygen diving. Hence, the maximum time of a single mission τ, using apparatus $oxy - CCR$ *AMPHORA SCUBA* and oxygen during a transit with excursions to a maximum depth of up to 15 mH_2O, should be set at $\tau \leq 150\,min$. In exceptional cases, this period should not exceed $\tau < 200\,min$.

vi. For the proposed technology of diving with the use of oxygen as a breathing medium with an allowable nitrox excursion, one type of premix[1] Nx 0.43 dosed at the level of $\dot{V} = 10.5\,dm^3 \cdot min^{-1}$ was used. This allows making oxygen dives with Nx excursion to the range of depths $[6,24]mH_2O$ and the maximum duration of the protective effect of diving apparatus during the deep excursion $\tau \leq 30\,min$ (Table A1.1).

TABLE A1.1

The maximum allowable exposure oxygen and N_x times.

Depth range	Maximum allowable excursion time	Remarks
[mH_2O]	[min]	
Oxygen		Maximum exposure time up to
6–12	10	240 min
12–15	5	After an oxygen excursion maximum exposure time decreases to 120 min
Nitrox: $0.43 O_2 / N_2$		The last 5 min is treated as an
6–24	30	emergency time needed to equalize potential delays

BASIC INFORMATION

1. During training dives the presence of a diving medical officer (doctor) trained in providing medical support during oxygen/Nx dives with the use of *oxy – CCR/Nx – SCR AMPHORA SCUBA* apparatus is recommended.
2. The listed rules on how to use the technology should be strictly adhered to in order to ensure divers' maximum safety.
3. Basic definitions applicable to the use of the technology are listed in Table A1.2.
4. The technology of oxygen diving is divided into two types: single exposure and transition procedures.
5. The transition procedure allows for oxygen/nitrox excursion when the *oxy – CCR/Nx – SCR AMPHORA SCUBA* apparatus is used (Table A1.1). The oxygen dives procedure can also be applied for diving with the use of *oxy – CCR Oxy – NG SCUBA*, *oxy – CCR FROG SCUBA* and *oxy – CCR CODE SCUBA* apparatuses, but the time of protective effect should be taken into consideration (Table A1.3).
6. Only trained divers who have successfully passed the oxygen tolerance test should be allowed to dive with oxygen and Nx 0.43.[2]
7. The Nx excursions described in this technology, which are part of the oxygen missions, do not require additional decompression procedures.
8. During the training, the night before the exposure diver should sleep at least 6 h. Other requirements for the diver's rest periods during the training are given in Table A1.4.
9. After a meal diver should rest minimum 2 h.
10. Oxygen diving procedures are less dangerous for a trained diver. To maintain combat readiness, it is necessary to run training dives in the military units.

    Each diver should dive at least four times a month with oxygen as a breathing medium and at a minimum time of swim [60, 80] min with the speed of about [0.4, 0.5] kn at the depth of [3, 6] mH_2O. He/she should also, once a month, exercise air diving at a depth of about 20 mH_2O with a minimum dive time of 20 min.

TABLE A1.2
The terms used in diving procedures using oxygen and N_x.

Specification	Definition
Transit with excursion limits	Diving to a maximum dive depth of 6 mH_2O or shallower allow the diver to make a brief excursion to depths as great as 15 mH_2O
Transit	The portion of the dive spent at a depth of 6 mH_2O or shallower
Excursion	The portion of the dive spent at a depth deeper than 6 mH_2O, but shallower than or equal to 15 mH_2O
Excursion time	The time between the diver's initial descent below 6 mH_2O and returning to 6 mH_2O
Single-depth limits procedure	In this procedure the diver is allowed to make a dive to a depth of 15 mH_2O
Repetitive oxygen diving	Diving with a closed circuit apparatus with oxygen as a breathing medium, when an interval between single dives is less than 2 h
Off-oxygen interval	The time from when the diver discontinues breathing oxygen on one dive until he/she begins breathing oxygen again from the closed circuit apparatus, having oxygen as a breathing medium, on the next dive

TABLE A1.3
Single-depth oxygen exposure limits.

Maximum diving depth	Maximum oxygen time
[mH_2O]	[min]
6	240
7	142
8	90
9	60
10	42
11	30
12	22
13	17
14	13
15	10

TABLE A1.4
Required diver rest time.

Full rest		Absence from work		Time of availability[†]
Before dive	After dive	Before dive	After dive	
[h]	[h]	[h]	[h]	[h]
2	2	2	2	2

† Time of diver's stay in the vicinity of decompression chamber after completion of dive.

If the break in oxygen diving is approximately one month, the diver should undergo an air dive training in a decompression chamber at a depth of approximately 20 mH_2O with a minimum dive time of 20 min.

If the break in diving is longer than a month, the diver should undergo initial additional air training at the approximate depth range of [10, 15] mH_2O with a minimum length of chamber dive of 20 min, after which he should undergo basic training on air before commencing oxygen dives.

Return to oxygen diving should take place with a gradual increase in difficulty of the performed tasks.

11. There should be at least a 2 h break between oxygen dives; otherwise, the next dive should be treated as a repetitive exposure. Repetitive exposures should be avoided.[3]

12. For the transit procedure the total time of oxygen exposures should not exceed 240 min, or 120 min if an excursion was made below 6 mH$_2$O. For the single exposure procedure, oxygen exposure times are shown in Table A1.3.

13. The surface interval between the end of one oxygen dive and the next oxygen dive with an Nx excursion should be at least 12 h.

14. To the total oxygen exposure time, Nx excursion dive time should not be added. But after an Nx excursion, the total oxygen exposure time should not exceed 120 min. Any accidental dive below the transit depth excludes the possibility of continuing the process of diving.

15. If after a series of three dives, between which surface intervals were shorter than 12 h and the dives are to be continued, an additional obligatory rest time of 12 h should be applied.

16. Oxygen dives at depths greater than 6 mH_2O should be avoided. They are allowed only in combat situations, during an exercise and training in training centers, where qualified personnel and medical equipment are available. Divers should be intensively and permanently trained in deep diving procedures in training centers so they can safely apply them during combat or emergency situations.[4]

DIVING

17. Before commencing a dive with oxygen as a breathing medium, the diver should perform a single purge of a breathing loop.[5] It involves closing the oxygen cylinder, sucking the breathing medium through the mouthpiece and blowing it out through the nose.[6] Then the cylinder should be opened and the breathing loop should be allowed to fill up with the fresh gas.[7]

The term oxygen dives as used in this instruction refers to dives those for which the oxygen content of the breathing medium exhaled by the diver exceeds $> 75\%_v O_2 / N_2$, so there is no need to purge intensively the breathing loop. It is recommended, however, that the oxygen content should stabilize at a low value of its permissible content.

18. When diving with the use of oxygen as a breathing medium the descent should be slow, at a rate of no more than $3\, mH_2O \cdot min^{-1}$.

19. While at the depth diver should not purge the apparatus' breathing loop.
20. When changing the breathing medium from oxygen to Nx there is no need to purge the breathing loop of the apparatus.
21. In a tactical situation, ascent should be slow enough for the excessive amount of breathing medium not to be emitted into water, maintaining the concealment of action, and such ascent should not be faster than the breathing medium bubbles.
22. In combat, when the dive is continued with oxygen as a breathing medium, after Nx excursion and when ascent to the surface is not possible, the breathing loop must be purged with oxygen at a depth ranging [15, 10]mH_2O,[8] maintaining secrecy of the action as far as possible.[9] The purge procedure of the apparatus' breathing loop should be performed as follows:

    • switch to oxygen
    • empty the breathing bag by pressing the apparatus to the diver's body, which will result in pressing the breathing bag
    • open the cylinder with oxygen
    • take one full breath[10]
    • empty the breathing bag by pressing the apparatus to the diver's body, which will result in emptying the breathing bag
    • close the relief valve

    During the initial stages of the training, switching to oxygen can be carried out at depths not greater than 7 mH_2O, with a single purge of the apparatus' breathing loop with oxygen.

    After completing the purge procedure of the apparatus' breathing loop, the diving can be continued with the use of oxygen as a breathing medium at depths up to 6 mH_2O.

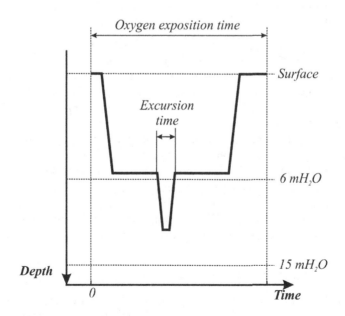

FIGURE A1.1 Example of the oxygen transit exposure.

23. The total time of each phase of the dive is calculated together with the change in depth (Figure A1.1).

24. In practice, maintaining secrecy of the operation with oxygen excursions is difficult to carry out due to the relatively short time of such oxygen excursions (Table A1.2, Table A1.3 and Figure A1.1). Therefore, nitrox excursions are preferred.

25. The following factors increase the risk of *CNSyn*:

    • hard work[11]
    • hypothermia or hyperthermia during the dive
    • repetitive dives
    • poor training or individual increased sensitivity to hyperbaric oxygen
    • poor mental condition of the diver

    The diver, who is exhausted, is stressed or demonstrates negative mental attitude is not allowed to dive. Strenuous exercise or total lack of activity, especially during dives at depths greater than 6 *mH$_2$O* with oxygen as breathing medium, shall be avoided. Repetitive dives shall be avoided unless they are part of the training or exercise or are a result of circumstances such as a combat or rescue mission.

26. The diver must abort the dive and start the ascent in the following cases:

    • at an order of the diving supervisor
    • at an order of the senior diver in the pair of divers
    • at a request of one of the divers in the pair
    • when symptoms of oxygen toxicity are noticed
    • fatigue and malaise
    • equipment malfunction
    • when the pressure of breathing medium in cylinder *p ≤ 5.0 MPa*

27. The diver should report to the senior diver of the pair of divers on his current well-being during the dive; every change in movement parameters of the pair of divers must be preceded by an agreed rope signal or other signal.

28. All the described procedures regarding diving with the use of oxygen as a breathing medium can be applied[12] to diving operations conducted in high-altitude reservoirs.

29. Air transport immediately after the dive with the use of oxygen as a breathing medium is prohibited only if the diving with the use of oxygen was part of a diving operation during which, at different depths, breathing mediums other than oxygen were used.[13]

    In such a case, the possibility of air transport should be considered for each operation individually. In other cases, almost immediate air transport after dive is possible up to 300 *m* altitude.

CENTRAL NERVOUS SYNDROME

30. During the experiments with the *central nervous syndrome* (*CNSyn*) oxygen poisoning, several symptoms were observed such as restlessness, pallor of the face, lips and eyelids trembling, nausea, cramps, dizziness, lack of

coordination, visual and auditory hallucinations, narrowing of the field of view[14] and slurred speech. These symptoms rarely precede a seizure. The beginning of generalized seizures is usually sudden. The attack begins with the tonic phase, usually lasting 30 s, during which the diver loses consciousness and stops breathing. This is followed by the clonic phase with uncoordinated movements of the entire body. The attack usually lasts approximately 2 min. While in water, the unconscious diver should be protected from drowning. The preferred procedure is to wait for the return of consciousness, even allowing the period of apnea to last up to 5 min. Immediate ascent of the poisoned diver can cause pulmonary barotrauma because poisoning is usually accompanied by airway obstruction due to contraction of the glottis.

MEDICAL RECOMPRESSION TABLES

31. During medical treatment, medical tables included in the appropriate instructions should be used.[15]
32. During medical treatment, inviolable oxygen and air[16] supply should be kept (Table A1.5).

TABLE A1.5
Calculation of necessary chamber reserve of oxygen and air for two divers.

OXYGEN

Specification	Number of divers	Time of stay	Pressure calculator	Lungs ventilation	Volume
		[min]	[atm atm⁻¹]	[dm³ min⁻¹]	[Mm³]
Treatment for	2	80	2.8	30	13.4
extended regime		40	2.8		6.7
TT 6A USN		30	2.5		4.5
		30	2.2		4.0
		120	1.9		13.7
		30	1.6		2.9
TOTAL					45.2

it makes 8 cylinders $40\,dm^3/15.0\,MPa$

AIR

Specification	Geometrical volume of chamber	Pressure calculator	Volume
	[m³]	[atm atm⁻¹]	[Mm³]
Treatment according to Table 11 MW	V_k	11	$11 \cdot V_k$
Required reserve 100%			$11 \cdot V_k$
TOTAL			$22 \cdot V_k$

TRANSIT PROCEDURE

33. Transit excursion procedure is the most common operation during special operations. An example of the profile of this procedure is shown in Figure A1.1.
34. If a task requires diving with only oxygen as a breathing medium[17] to a depth of more than 6 mH_2O for the time longer than the permissible excursion time, the procedure of single exposure should be applied.
35. The diver, who is at a depth of 6 mH_2O or less, can make an oxygen excursion to a greater depth, but during planning phase he/she should calculate the dive following the rules below:

   • maximum dive time[18] should not exceed 240 *min* without taking any excursions. After one excursion, the time is shortened to 120 *min*, excluding the excursion time; however, the protective time of the absorber canister of *oxy – CCR/Nx – SCR AMPHORA SCUBA* apparatus limits the maximum oxygen exposure time to 150 *min*
   • a single oxygen excursion can be started at any time while at a depth of transit, provided that the transit time does not exceed 120 *min*
   • the diver must return from an excursion to a depth of 6 mH_2O or less, finishing the excursion, before the end of the time permissible for such excursion (Table A1.2)
   • only one oxygen excursion is allowed to a depth exceeding 6 mH_2O and not exceeding 15 mH_2O
   • the time planned for an excursion is limited due to the maximum depth reached during the excursion (Table A1.2)

36. If during the oxygen diving an accidental oxygen excursion happens, the following procedure should be applied:

   • if the depth and/or time of the excursion exceeds the permissible time limits or is a second excursion, the dive should be aborted and the divers should return to the surface
   • if an accidental excursion is the first one and has not exceeded the permissible limits, diving can be continued until all the relevant limits and/or the allowable depths are reached; any additional excursions to a depth greater than 6 mH_2O should not be undertaken[19]
   • if an accidental excursion has exceeded the limits laid down for the transit procedure and the dive can be regarded as a single dive to the maximum allowable depth, the procedure of single exposure can be used and the dive can be continued in accordance with this procedure
   • if the diver after an uncontrolled descent is not sure how long he/she stayed at a depth of less than 6 mH_2O, that dive should be aborted

37. The diver, who is at a depth of 6 mH_2O or less, can make an Nx excursion to a depth 24 mH_2O with a maximum duration of 30 *min*.
38. It is not allowed to make Nx excursions if the diver is in the process of oxygen excursion.

SINGLE EXPOSURE PROCEDURE

39. Table A1.3 shows acceptable exposure time limits applicable to dives using oxygen as a breathing medium for single exposure procedure (Figure A1.2). In contrast to the transit procedure, during the single exposure procedure additional excursions are not allowed.
40. The diver does not have to spend the whole time of exposure at the same depth, but the allowed exposure time is limited since it is a function of the maximum depth reached during the dive.
41. When using a single exposure procedure, Nx excursions are not allowed.

EXCEPTIONAL EXPOSURES

42. During a mission there may arise exceptional situations, which include repetitive dives using oxygen as a breathing medium. These are oxygen dives conducted after intervals shorter than 2 h after the end of the previous oxygen dive. Such dives can only be conducted in emergency situations.
43. If a repetitive dive is planned with oxygen as a breathing medium, the effects of a previous dive on permissible time and depth of the planned repetitive dive should be taken into consideration (Table A1.6).

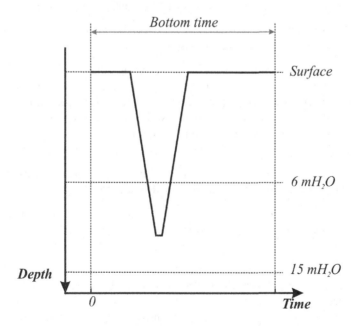

FIGURE A1.2 Example of the oxygen single exposure.

TABLE A1.6
Methods for calculating the permissible oxygen exposure limits for successive dives.

Specification	Method of calculating maximum successive dive time	Excursion to a greater depth
Transit with excursion limits	Subtract time on previous exposures from 240 *min* and the time left is the permitted time for successive dive	Allowed if such an excursion was not made on previous dive and the transit time did not exceed 120 *min*
If any of the described situations can be applied, the diver has to wait at least 2 *h* before his next dive		
Single-depth oxygen exposure limits	1. Determine maximum oxygen dive time for deepest depth. Diving depth is selected after analyzing previous and planned dives. 2. Subtract oxygen time on previous dives from maximum successive dive time defined in pt. 1 above	No excursion allowed when using single-depth limits to calculate remaining oxygen time
If any of the described situations can be applied to the single-depth limits, the diver has to wait at least 2 *h* before his/her next dive		

44. If the interval between the previous and planned oxygen exposure is greater than 2 *h*, the previous dive can be disregarded. However, it is recommended that such operations be planned only for exceptional situations during combat and during training in centers prepared for this type of exercise. We do not recommend use of these procedures in military units during standard exercises.

    If during an oxygen mission an *Nx* excursion was undertaken, the next dive cannot be made earlier than after 12 *h*.

45. Another unique situation encountered during combat missions is diving with the use of oxygen as a breathing medium after other types of exposure. There is no simple procedure that would allow calculating the permissible limits for oxygen exposures after a dive when air or breathing mixtures were used as a breathing medium.

    If the dives preceding the oxygen exposure led to the diver being exposed to a partial pressure of oxygen $\geq 0.1\,MPa$, the effect of a previous dive must be taken into account when planning the dive with the use of oxygen.

46. The interval between the dives is counted from the moment when the diver stops breathing the previous breathing medium until he/she begins to breathe oxygen.

    If the diver uses an oxygen closed circuit apparatus to carry out one part of diving and another apparatus with other than oxygen breathing medium in further diving, only the portion of the dive when the diver breathes pure oxygen should be counted as the oxygen exposure time.[20] The use of different types of diving apparatuses to perform special operations is allowed when proven technologies developed individually for each scenario are used.

EXAMPLES OF DIVE PLANNING

1. TRANSIT PROCEDURE

During the mission the divers remained at a depth of 6 mH_2O for 45 min and then dived to a depth of approximately 11 mH_2O. Design the plan to continue the dive.

As long as the divers have not exceed the maximum allowed depth of excursion $H_{max} \leq 12 mH_2O$, they can use the maximum time limit permitted for an excursion to a depth ranged [6, 12]mH_2O, equal to 10 min – Table A1.2. This time should be sufficient for divers to descent from the depth of 6 mH_2O, remain at the depth[21] and return to a depth of 6 mH_2O or less – Figure A1.1. Total time of transit and excursion will be maximum 45 + 10 = 55 min. Therefore divers can still remain within the depths ranged [0, 6]mH_2O for about 120 – 55 = 65 min – Table A1.2.

2. ACCIDENTAL EXCEEDING OF LIMITS DURING THE TRANSIT PROCEDURE

A pair of divers using oxygen as a breathing medium have noticed a malfunction of the compass and, busy with analyzing the situation, they figured out after 35 min of the dive that their depth gauge shows 16.7 mH_2O. Describe the procedure to be applied.

Since the depth of 16.7 mH_2O is greater than the maximum allowed for diving with the use of oxygen as a breathing medium for both transit and single exposure procedures, the divers should ascend immediately, aborting the dive.

3. ACCIDENTAL EXCEEDING OF LIMITS DURING THE TRANSITION PROCEDURE

A diver using oxygen as a breathing medium and using compass readings for direction suddenly noticed that his/her current diving depth is 9.7 mH_2O. He/she remembered, however, that five minutes ago he/she was at a depth of 5.4 mH_2O and all the time, until he/she noticed changes in depth he/she was sure that he/she did not exceed a depth of 6 mH_2O. Give instructions on how to proceed.

A diver made an unplanned excursion not exceeding 12 mH_2O. In addition, he/she probably did not exceed the allowed time for excursion, but for security reasons he/she should ascend to a depth not exceeding 6 mH_2O. He/she may continue diving at a depth of 6 mH_2O or less. He/she should remember that he/she must not take any additional excursions below a depth of 6 mH_2O until the end of the dive and that the maximum total diving time allowed for the transit procedure is now shortened to 120 min – Table A1.2.

The diver should not consider switching to the single exposure procedure because he/she is not certain how long and how deep he/she was below the depth of the transit.

4. PERMISSIBLE EXCURSION WITH OXYGEN AS A BREATHING MEDIUM

After 22 min swimming on compass reading at a depth of 4 mH_2O, a pair of divers had to go down to a depth of 8.5 mH_2O to avoid propellers of a surface vessel passing above them. They returned to the previous depth after 8 min. How should they further proceed?

After swimming, the divers have two choices when calculating the remaining dive time.

They can continue diving, not exceeding the depth of 6 mH$_2$O, and treat the time spent below 6 mH$_2$O as a permissible excursion. Since the total dive time is now shortened to 120 min, they have now 120 − 22 − 8 = 90 min of diving left – Table A1.2.

Optionally, according to the single exposure procedure, they can consider the current mission as a single dive to a depth closest to the next largest from the table of oxygen-permissible exposures[22]– Table A1.2. They will still have 60 − 22 − 8 = 30 min dive left, but they still can make further excursions, the depth of which should not exceed 10 mH$_2$O with the time of stay 42 − 22 − 8 = 12 min. For a depth of 11 mH$_2$O, the time of single oxygen exposure has been already exceeded.[23]

5. REPETITIVE DIVING FOR TRANSIT OPERATIONS

Divers must make a second dive using the transit procedure 90 *min* after the end of the previous dive to a maximum depth of 5.8 *mH$_2$O* with the 75 *min* length of stay and using oxygen as a breathing factor. How should they plan their dive?

The considered second dive will be a repetitive dive, since the interval between breathing with oxygen lasted less than 2 h. The permissible time of repetitive dive has to be calculated in accordance with Table A1.6. The maximum time of dive defined in this way is 240 − 75 = 165 min[24] or 120 − 75 = 35 min, if an excursion is taken to a depth in the range of [6, 15] mH$_2$O.

6. REPETITIVE DIVING FOR SINGLE EXPOSURE OPERATION

After completing a dive to a maximum depth of 8.5 *mH$_2$O* with the time of stay of 45 *min* and using oxygen as a breathing medium, a pair of divers undertake a second dive using the same equipment to a maximum depth of 7.6 *mH$_2$O*. Plan for this dive.

The maximum time allowed for repetitive diving according to the procedure of single oxygen exposure should be taken into consideration for the greater depth of the two dive profiles. For a depth of 9 mH$_2$O, exposure time is 60 min – Table A1.3. Then the time of the first oxygen exposure has to be subtracted, in this case 45 min, from the permissible time limit the selected maximum depth. The dive time calculated in this way is 60 − 45 = 15 min.[25] The total time of 60 min does not exceed the maximum duration of the protective effect of the absorber τ < 150 min, but if it is possible, the old absorbent should be replaced. Note that divers cannot perform any additional excursions to a depth greater than 9 mH$_2$O.

7. COMBINED DIVING

Divers are transported to the vicinity of the port using an underwater vehicle. Divers will breathe compressed air for a period of transport. The depth and time of transfer during air breathing will require zero decompression time according to the applied air decompression tables. After the divers are moved to the area near the port, they will leave the vehicle, descend to a depth of 6 *mH$_2$O* and switch to breathing oxygen in a closed circuit system. They will make a dive following compass indications. Plan the overall oxygen transfer.

The planned procedure will be a transit with one excursion to a greater depth using a closed circuit breathing apparatus with oxygen as a breathing medium. From this point of view, the previous diving with air as a breathing medium may be disregarded.[26] Therefore, it is not taken into account in planning a dive with oxygen as a breathing factor.

8. Single Exposure Operation

Divers after beginning the mission had to submerge to a depth of 7.5 mH_2O and stay there for 40 *min*, and after that they were to continue diving at a depth of 5 mH_2O. Define the plan for continuation of the dive.

The divers have exceeded the time of 10 min at a depth ranged [6, 12] mH_2O, so they cannot apply the transit procedure. As long as the divers do not exceed a maximum depth of $H_{max} \leq 8\,mH_2O$ they can use the maximum time available for a single exposure dive to a depth in the range of [0, 8] mH_2O, which is 90 min –Table A1.3. Given the fact that the divers remained at a depth of 7.5 mH_2O for 40 min, they have 90 – 40 = 50 min to remain at a depth of 8 mH_2O, or 60 – 40 = 20 min to remain at a depth of 9 mH_2O or 42 – 40 = 2 min to stay at a depth of up to 10 mH_2O – Table A1.3.

9. Single Exposure Operation

Divers after the beginning of the mission had to submerge to a depth of 8 mH_2O where they stayed 20 *min*. Design plans for continuation of the dive.

The divers have exceeded a period of 10 min at depths ranged [6, 12] mH_2O, thus they cannot apply the transit-type procedure. Since the divers do not exceeded the maximum depth of $H_{max} \leq 9\,mH_2O$ they can use the maximum time available for the single exposure dive to a depth of 9 mH_2O, which is 60 min or to a depth of 10 mH_2O, which is 45 min. Therefore, the divers can still remain in the water at a depth of 9 mH_2O for 40 min or they can make a dive to a depth of up to 10 mH_2O for a period of 22 min. Ultimately they can dive to a depth of 11 mH_2O for 10 min, but this may be useful only in special situations. Dives to 12 mH_2O for 2 min are not realistic – Table A1.3.

NOTES

1 *Nx* (43.0 ± 0.5) $\%_v\,O_2/N_2$; metering $\dot{V} = \left(10.5 \pm 1.0\right) dm^3$.
2 *Nx* (43.0 ± 0.5) $\%_v\,O_2/N_2$; metering $\dot{V} = \left(10.5 \pm 1.0\right) dm^3$.
3 Repetitive dives.
4 For example, lifesaving.
5 It is recommended that such activities be performed when preparing the apparatus before diving, and leaving it sealed until the beginning of the dive, but it does not replace the mandatory purging of the breathing loop directly before the dive, since such a procedure also applies to flushing the diver's respiratory system.

  6 Or by mouth after closing the mouthpiece.

  7 Through this, one should not breathe from the apparatus.

  8 After returning from an *Nx* excursion, it is recommended that the breathing loop be purged near the surface, checking first that this does not cause the diver to be detected.

  9 Purging should be performed in such a way that there is no excessive, single emission of a breathing medium into the water.

10 Inhale and exhale.

11 Oxygen diving should have small loads – swimming at speeds not exceeding 0.5 *kn*.

12 Without any modifications.

13 For example, air, nitrox (*Nx*), heliox (*Hx*), trimix (*Tx*), etc.

14 Known as tunnel vision.

15 For example, STANAG 1432.

16 Reserve applies primarily to the treatment of decompression sickness resulting from the *Nx* excursion.

17 If an excursion using *Nx* as a breathing medium is possible, it is preferable from the physiological point of view but is not always justified from a tactical point of view.

18 Time of breathing with oxygen.

19 Unless the oxygen profiles are acceptable for the single exposure procedure.

20 This happens when diving is not too deep and does not last too long.

21 maximum to 12 *mH$_2$O*.

22 For the depth of 9 *mH$_2$O* the permissible time of diving is 60 *min* – Table A1.3.

23 At the depth of 11 *mH$_2$O*, permissible exposure time is up to 30 *min*.

24 240 *min* is the maximum allowed oxygen exposure time for the transit procedure, without the excursion made to a depth of more than 6 *mH$_2$O*.

25 60 *min* for any exposure permissible for 9 *mH$_2$O* minus 45 *min* of first exposure.

26 An additional reason to neglect the air exposure is the fact that it required zero decompression.

Appendix 2
Oxygen Tolerance Test

PRESSURE AND OXYGEN TOLERANCE TEST

1. It is recommended that all divers using oxygen or artificial breathing media be screened for their tolerance to hyperbaric oxygen by taking an oxygen tolerance test (OTT).
   - both diving candidates and divers and should take a hyperbaric pressure test and oxygen tolerance test
   - a pressure test should be administered as often as possible and should be obligatory for divers seeking to obtain a higher qualification and prior to a periodic medical examination that pronounces a diver fit or unfit for service
2. An oxygen tolerance test should be performed twice with at least a one-week interval. The procedures for the aforementioned tests have been tested and verified in practice. It should be borne in mind that these tests are subject to some specific risks and should therefore be performed with particular care.

OBJECTIVES

3. A pressure test provides answers to the following questions:
   - Can a diver, when necessary, make a fast descent?
   - What is the functional status of his/her paranasal sinuses and Eustachian tube?
4. An oxygen tolerance test is used to discover the following:
   - whether a diver, in case of necessity, can undergo a hyperbaric chamber treatment with oxygen as a breathing medium
   - whether a diver, due to his/her individual resistance to the toxic effects of oxygen on the central nervous system, can safely dive using a special diving apparatus that employs pure oxygen as a breathing medium
   - whether a diver, due to his/her individual resistance to the toxic effects of oxygen on the central nervous system, can safely dive using a gas mixture as a breathing medium in which oxygen partial pressure is higher than 100 *kPa*
   - The suitability of a diver and the likelihood of convincing him/her that, regardless of the test result, he/she is providing important information to personnel involved in the treatment of divers in case of an accident or illness related to diving.[1]
5. Administering an oxygen tolerance test should be confirmed by an entry in the diver's logbook. The right to make such an entry rests with an authorized physician who executes the test. The entry must be made regardless of the result of the test.

INTRODUCTORY ISSUES

6. Passing the test requires the diver has sufficient theoretical knowledge. Prior to the test he/she should review his/her knowledge of physiopathology of diving, especially the issues concerning the toxic effect of oxygen on the central nervous system and lung parenchyma (Table A2.1).

7. Before the *OTT*, the diver should take a written mastery test of the material referred to in point 6. The test results should be kept by the doctor or should be stored in the unit conducting the oxygen tolerance test.

8. The hyperbaric chamber adapted to perform pressure and oxygen tolerance tests should consist of two compartments, or at least of a single compartment with a wet porch used to lock out divers and equipped with the following:

   • an oxygen concentrator with hermetic oxygen circulation[2]
   • a fire suppression system inside the chamber, an extinguisher approved for use in a hyperbaric environment, or extinguishing media inside the chamber[3]
   • emergency breathing systems – the number dependent on the nominal operating capacity[4] of the chamber – to be used in case the atmosphere in the chamber[5] is contaminated

TABLE A2.1
Topics to study before an oxygen tolerance test and pressure test.

Topics	Topics for self-study (to be verified)	Issues to be covered under the specific topics
Organization of classes and regulations applicable to tests (seminars & lectures)	(a) Diving regulations (b) Procedures applicable to diving activities in the Polish Navy	(a) Regulations and rules applicable to hyperbaric tests (b) Organization of classes (c) General occupational Health and Safety regulations
Physiopathology of diving – revision (seminar)	(a) An overview of human anatomy and physiology (human cells and tissues, body organs, general structure and physiology of cardiovascular system, overall structure and physiology of respiratory system, structure and physiology of hearing, nasal sinus) (b) Accidents and diving-related illnesses (eye squeeze, ear squeezes, sinus squeeze, pulmonary barotrauma, diver's blowout, diver's fall, nitrogen narcosis, asphyxia, carbon monoxide or carbon dioxide poisoning, poisoning with fumes, oxygen poisoning, body overcooling and overheating) and procedures applicable to diving accidents	(a) Oxygen pulmonary toxicity (b) Toxic effect of oxygen on central nervous system

- a gas composition monitoring system that measures oxygen and carbon dioxide content in the atmosphere inside the chamber
- wired or wireless electric signal buttons to signal an emergency situation, one for each diver present in the hyperbaric chamber connected in such a way that they can be easily held in hand during the entire test. They are installed in addition to a wooden or rubber hammer supplied for each diver as an emergency signaling system. In the absence of electric signal buttons one wooden or rubber hammer per person can be used.
- an internal first aid kit
- a technical monitoring system[6] or such an arrangement of portholes that allows accurate observation of people in the chamber
- an electronic communication systems, main and emergency,[7] together with emergency power supply in case of loss of power to ensuring operation of the communication equipment for at least 12 h
- an internal lighting system with emergency power system, ensuring continuous operation for at least 12 h
- an airlock for passing small items to and from the hyperbaric chamber[8]
- an autonomous heating system if the chamber is located in an unheated room

9. A candidate for testing must have a current medical certificate issued by the relevant military medical board, confirming his/her fitness for service as a diver. Prior to commencing the tests, the diver must have taken sufficient rest. On the day the tests are to be carried out, he/she should not perform any rigorous physical exercises.
10. A candidate for a pressure test and for an oxygen tolerance test should undergo a medical examination prior to undertaking these tests. An example of the medical examination form required to be filled out – it is shown in Table A2.2.
11. Divers and the physician supervising the hyperbaric tests must undergo similar medical examination in accordance with Table A2.2.
12. The diver undergoing the tests should be familiar with the rules of conduct inside the hyperbaric chamber and operating the chamber systems, especially the internal oxygen system. The fact of completing the required training should be confirmed by the relevant statement:

> *Hereby I confirm that I have been informed of the risk of oxygen toxicity. I agree to undergo the pressure test* and oxygen tolerance test* in accordance with the regulations in force applicable the standard pressure and oxygen tolerance tests.*
>
> *I express my consent to recording the test* and subsequently to using it later for training purposes without revealing my personal details.**
> ** delete as appropriate*

13. Tests should be supervised by either diving physicians or other trained personnel who are able to provide medical assistance under hyperbaric conditions. A standby diver providing emergency support for the pressure

TABLE A2.2
Example of medical examination and equipment test form

Medical examination and equipment test

To be completed by the tested diver	To be completed by pressure test safety diver
Name:	Name:
Surname:	Surname:

Yes	No	Enquiry	Yes	No	Enquiry
		Are you able to dive?			Are you able to dive?
		Have you consumed alcohol in the preceding 24 *h*?			Have you consumed alcohol in the preceding 24 *h*?
		Are you rested?			Are you rested?
		Have you tested your equipment for proper operation and checked its completeness?			Have you tested your equipment and checked its completeness?

Medical exam, date:	Medical exam, date:
Oxygen tolerance test date:	Oxygen tolerance test date:
Last dive	Last dive
#date:	#date:
#breathing medium:	#breathing medium:
#depth:	#depth:
#bottom time:	#bottom time:
I confirm the above data	I confirm the above data
Signature of diver:	Signature of diver:
Medical examination (to be completed by the doctor)	
Pulse	Pulse
Arterial blood pressure	Arterial blood pressure
Number of breaths per minute	Number of breaths per minutes
Doctor's opinion:	Doctor's opinion:
Date, signature and the seal of the doctor:	Date, signature and the seal of the doctor:

test should be specially trained and possess skills required to undergo a fast compression.

The basic document applicable to organization, conduct and safety rules during dives, including simulated hyperbaric exposures in pressure chambers, is the Diving Regulations.

The regulations specify training standards for physical endurance for pressure conditions applicable to all divers, and particularly personnel supervising the tests in the hyperbaric environment, subsequently allows them to efficiently operate when the tests are being performed.

It is anticipated that such training should be carried out at least twice a month up to the maximum allowable dive working pressure adopted for a given group of divers. As regards the personnel supervising hyperbaric tests, this should be the maximum allowable pressure[9] for using air as a breathing medium in a simulated dive profile, i.e. the pressure equivalent to a depth of 60 mH_2O.

This is because the divers who are employed to provide necessary assistance in support of not only hyperbaric tests but also routine training attended by teams of other divers have to be in good condition. They also must be available when needed to act as attendant (standby) divers in standard hyperbaric treatment procedures in the event of accidents or divers' diseases. Some of these procedures require, in the initial phase of treatment, pressurizing the sick diver and commencing recompression from the maximum treatment pressure corresponding to a depth of 50 mH_2O.[10]

At the same time it is intended that each training to the maximum working pressure[11] is preceded by an interval of one day, by training to half of the required maximum pressure.[12]

When a diver candidate begins training as a standby diver and when intervals in diving/training longer than one month are not caused by his/her sickness, the training will cover a cycle of four pressure trainings conducted at one-day intervals. Two of them will be held at a pressure equivalent to a depth of 10 mH_2O while for the other two it will be equivalent to 1/2 and to the maximum working pressure[13] respectively.

RULES OF CONDUCT DURING TESTING

14. First, the tested diver and the attendant diver undergo the pressure test. They are in the decompression chamber that is then pressurized, using air, to the simulated dive depth of 30 mH_2O, at the maximum rate that can be tolerated by the tested diver but not higher than $20\,mH_2O \cdot min^{-1}$. The attendant diver should be carefully selected and trained for this type of test.

15. After reaching a depth of 30 mH_2O, the hyperbaric chamber should be intensively ventilated for 1 min in order to reduce the internal temperature.

16. After the ventilation, the pressure in the hyperbaric chamber should be reduced to the simulated diving depth of 18 mH_2O at a rate of $10\,mH_2O \cdot min^{-1}$. The maximum emergency pressure reduction rate[14] is $18\,mH_2O \cdot min^{-1}$.

17. After reaching a depth of 18 mH_2O, the attendant diver should be replaced with a second attendant diver. To achieve this, the second diver should be compressed in a separate compartment of the hyperbaric chamber to the

‌

simulated depth of 18 mH_2O at a compression rate of $[6,8] mH_2O \cdot min^{-1}$. After reaching this depth, the second attendant diver and the tested diver move to the compartment in which the oxygen tolerance test is to be carried out, while the first attendant diver moves to the compartment in which the second diver was pressurized. After sealing the compartments, the first attendant diver is decompressed in accordance with the air decompression table. It is recommended that extended decompression time be applied due to the drastic conditions of compression.

> The decompression time should also be additionally extended if, apart from the high compression rate, other unfavorable conditions arose that could have affected the safety of decompression. Even if the no-decompression profile is chosen for the first attendant diver, for safety reasons, he/she should remain in the decompression chamber for [1, 3] *min* at 3 mH_2O. If an oxygen decompression is planned for the attendant diver, he/she must have the oxygen mask removed from his/her head and, only holding it with his/her hand, press the mask to the face making sure it is tight-fitting and that in case of oxygen toxicity or loss of consciousness it freely falls off his/her face.

18. After replacing the attendant divers, the tested diver should be again briefly instructed on how to use the inhaler oxygen masks. Then he/she should begin the test, which involves breathing oxygen, in a horizontal position, through the tight-fitting mask for 30 *min* without any pauses in breathing.[15]
19. The attendant diver should, at all times, observe the tested diver and if there are any signs of hypersensitivity related to hyperbaric oxygen he/she should terminate the tested diver's breathing from an oxygen inhaler.[16]

> The most often mentioned symptoms of oxygen poisoning are seizures. It was noticed that they do not occur immediately after an exposure to oxygen at a pressure lower than 300 *kPa*. Less specific symptoms were also observed, such as restlessness, pale face, trembling lips and eyelids, nausea, cramps, dizziness, impaired coordination, visual and auditory hallucinations, narrowing of the field of vision[17] and speech problems. During the oxygen tolerance test, these symptoms rarely precede seizures.
>
> An attack of oxygen convulsions occurs without warning. It begins with the tonic phase lasting approximately 30 *s*, during which the diver loses consciousness and due to a hang-up of the respiratory system the respiration stops. This is followed by a clonic phase in the form of seizures with uncoordinated movements of the entire body. The whole attack usually lasts approximately 2 *min*, but apnea may last longer. Safe period of apnea during an attack may take up to [5, 8] *min*.
>
> One of the methods of diagnosing a diver's resistance to oxygen's toxic effect on the central nervous system is observation of respiratory actions before and during the test. It is believed that if during the examination before the test the diver takes fewer than six breaths per minute, he/she can be considered vulnerable to oxygen toxicity.

Similarly, if during the examination, breathing shows a tendency to drop below four breaths per minute, the appearance of oxygen poisoning should also be expected.

20. During the oxygen tolerance test, for security reasons, the hyperbaric chamber should be ventilated to prevent the concentration of oxygen in the chamber above 25%,. The partial pressure of carbon dioxide in the chamber atmosphere should not exceed 1 kPa.

21. When monitoring the breathing atmosphere, it should be remembered that oxygen and carbon dioxide are heavier than nitrogen, so if the atmosphere in the chamber remains stable, those gases will tend to concentrate in the lower parts of the chamber. Samples for analysis should be taken from or sensors should be located in those areas.

    If the hyperbaric chamber is not equipped with a monitoring system, it cannot be used for hyperbaric tests.

22. When the test driver has breathed oxygen for 30 min, decompression to the surface should start, at a maximum speed not exceeding $18\,mH_2O \cdot min^{-1}$. Regarding the health safety of the attendant diver, the recommended rate of decompression should not exceed $10\,mH_2O \cdot min^{-1}$. Despite the fact that no decompression is planned if the test runs trouble-free, a 1 min decompression stop at a depth of 3 mH_2O should be made.

FINAL PROVISIONS

23. Attention should be paid to ensuring attendant diver safety, employing the necessary number of them in the test.

24. During a pressure test the tested diver should, at all times, provide information on his/her current condition and the patency of Eustachian tubes and sinuses during changes in pressure.

    The technical support personnel should immediately stop the compression if any of the divers raises an alarm. For this purpose, it is recommended that a wired or wireless button be used to start the buzzer,[18] and a wooden or rubber hammer should be used as an emergency measure to initiate an alarm. In the absence of electrical signaling, each person should be provided with one wooden or rubber hammer – Table A2.3. The divers should also be monitored by using technical means of observation.[19] In the event of absence of such means, an observer should be placed by a porthole so that accurate observation of divers can be performed. All applicable signals should be discussed in advance during the briefing with the divers.

    Signals should be discussed during the pre-dive briefing. The suggested signals are given in Table A2.3. In the case of communications system failure, emergency signals are used to conduct simple communications. There should be a foil-covered table, showing a list of signals prescribed for use, in each compartment of the decompression chamber and on a control desk. During hyperbaric tests, each diver taking part should have their own copy of the table of signals.

TABLE A2.3
Set of acoustic signals used in case of failure of communications system in decompression chambers.

No	Signal	Graf	Meaning of signal during communication to decompression chamber	Meaning of signal during communication from decompression chamber
1.	Single signal	●	How do you feel?	I am OK, I understood
2.	Double signal	●●	I am raising pressure	Raise the pressure
3.	One single signal and one double signal	● ●●	Start breathing from the breathing system	I started breathing from a breathing system
4.	Three signals in a row	●●●	I am reducing the pressure	Reduce the pressure
5.	Number of signals (at least six) In a group	●●●●	—	Stop changes in pressure/stop
6.	One single knock and 4 in a group	● ●●●●	Open the lock	Open the lock
7.	Two double knocks	●● ●●	Stop breathing from the breathing system	I stopped breathing from the breathing system

If signal 1 is sent from the chamber following signal 1 sent to the chamber, it means, "I am OK." If signal 1 is sent from the chamber following signals 2, 3, 4 or 7 sent to the chamber, it means, "I understand." After confirmation, the command sent to the chamber should be repeated, for example:

Command sent to chamber	Reply on hearing the command	Reply on executing the command
● ●●	● ● ●●	● ●●

25. Upon its successful completion, the first test of the series of oxygen tolerance tests should be repeated after a week. Only upon a successful completion of the second test can an entry confirming successful completion of the oxygen tolerance test be made in the diver's personal logbook.
26. The air pressure test should be repeated on several occasions, e.g. before a periodic medical examination confirming fitness for service as a diver; when exams for higher diving qualifications or extended authorizations are taken; when the diver is required to confirm his/her qualification or authorization; during training to improve diving skills conducted outside his/her unit in his/her home military unit prior to any diving operations[20] scheduled to be performed in the near future.
27. Pressure tests can be done only upon approval by a duly authorized physician from the diving medical team and only in his/her presence.
28. If during the oxygen tolerance test[21] the diver suffers an attack of oxygen convulsions, the test is finished with a negative result and should not be repeated, and the diver receives in his/her logbook an entry that disqualifies him/her

from diving on a breathing media other than air. In such case, the depth of the dive using air as a breathing medium should be limited to 40 mH_2O.

29. If during the oxygen tolerance test the diver developed symptoms of malaise and his/her breathing rate dropped to four or fewer breaths per minute or symptoms of oxygen poisoning other than those of convulsion occurred, the test should be aborted and repeated after one week. If, again, the tested diver experiences any disorders during the repeated test, the applicable procedures are the same as in the case of oxygen convulsions. If during the second test the diver shows no symptoms of poisoning, he/she is considered to have successfully completed the first stage of the oxygen tolerance test, and the next test is carried out after one week. If during the second test[22] no problems are recorded, an entry confirming the successful completion of the oxygen tolerance test is made in his/her diving log. If, however, during the third test diver shows even the slightest symptoms of oxygen toxicity, he/she should be considered to have experienced oxygen convulsions.

   If the diver passed the first test successfully, and during the second test disorder symptoms or symptoms of oxygen toxicity other than oxygen convulsions occur, the test should be repeated twice at one-week intervals. If during the repeated tests the slightest symptoms of oxygen toxicity occur, the diver should be treated as if he/she experienced oxygen convulsions.

30. After each test, the tested diver should stay at least 1 h near the decompression chamber – the time to reach the chamber should not exceed 5 min. If this is the only decompression chamber at the test site, it should not be used at all for 30 min after the end of exposure and should only be available for a potential treatment of the test diver.

31. Upon completion of the oxygen tolerance test, the pressure test or both of the tests at the same time, flying is not permitted until 12 h from the moment the diver left the decompression chamber.

32. When it is necessary for the diver who has completed the oxygen tolerance test to undergo medical treatment, the pulmonary oxygen toxic dose inhaled should be taken into account. It can be assumed that two hours after the end of the test the diver's organism is completely regenerated owing to the oxygen toxic effect of on the lung parenchyma.

33. Special precautions should be taken to prevent accidents and diseases among the personnel supervising hyperbaric tests.[23] They are also obliged:

   • to remain in the vicinity of the hyperbaric chamber for an hour after exposure
   • to observe the rule that forbids air transport for 24 h
   • to hold a valid certificate confirming fitness for serve as diver, issued by a relevant authority
   • to attend an appropriate training designed to adapt their bodies to fast compression conditions.
   • to complete a relevant diving training

34. The training mentioned in the last point is not required if the diver partici-
    pates in pressure and oxygen tolerance tests as an attendant diver at least
    twice a week. It is also recommended that the medical personnel be exam-
    ined before they start performing their duties in support of the dives.

NOTES

1 The possibility of undertaking the treatment using oxygen or air.
2 No emission of oxygen into the chamber.
3 Such as water, sand, etc.
4 One for each diver in the hyperbaric chamber.
5 For example, in case of fire.
6 Internal TV system.
7 Communications if the main line is damaged.
8 If the chamber does not have sanitary facilities, the lock should be adapted to pass
   through faces in the sealed container, and such container should be a part of chamber
   equipment.
9 Together with personnel supervising tests.
10 For example, medical Treatment Table *TT 6A USN*.
11 To a depth equivalent to 60 *mH$_2$O*.
12 To a depth equivalent to 30 *mH$_2$O*.
13 To a depth equivalent to 30 *mH$_2$O* and 60 *mH$_2$O*.
14 For example, to abort the test due to diver's malaise.
15 If the diver does not show symptoms of oxygen poisoning.
16 The test diver should as soon as possible start breathing air from the hyperbaric
   chamber.
17 Also called tunnel vision.
18 Bell, buzzer or other acoustic signal.
19 Cameras.
20 For example, before dispatch for sea diving cycle to medium and large depth, or the
   execution of major underwater jobs.
21 First or second.
22 The third test, in turn.
23 Attendant divers.

Appendix 3
Factor Analysis

In the classical approach, efforts were made to find algebraic models[1] that were approximate deterministic mathematical descriptions[2] of the relationships between the magnitudes[3] defining an object of study[4] by using, known since the 18th-century paradigm[5] of *Leonard Euler*, which states that most physical phenomena can be modelled with differential equations.

In engineering applications, semi-empirical methods are often used to find algebraic models. In contrast to classical models, in semi-empirical models not all factors or parameters of an algebraic equation have physical interpretation.

Today also increased interest in empirical models is observed. They are usually developed using *exploratory data analysis (EDA)*, also called *data mining.*[6]

In empirical models, factors and parameters of algebraic equations have no physical interpretation. Such a model is a function of best empirical data approximation. The next step in the development of empirical modelling is the use of *neural networks*, in which an algebraic equation cannot be identified in the plain form. In the case of neural networks, the formula is contained in the interaction between layers, numbers and levels of neurons modelled by computer software.

All of these models must be validated, optimized, or even made on the basis of experimental data (Figure A3.1).

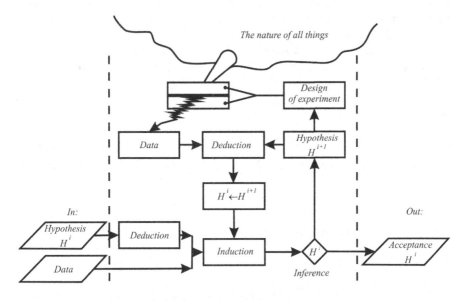

FIGURE A3.1 Diagram of the optimization process of models for studying objects of knowledge and thus learning more about them on the basis of planning further experiments.

INTRODUCTION

One of the methods of data mining is the *factor analysis* (*FA*). The concept of factor analysis involves three methods, identified as *factorial analysis, factor analysis* and *component analysis* (Bastlevsky A.T., 2010; Hinkelmann K., Kempthorne O., 2005). The first one concerns the planning of an experiment, the other two relate to replacing variables with a smaller number.[7]

The notion of *FA* includes the statistical methods considered here, which will be used to replace independent variables describing an object[8] with a set of new variables, more convenient for practical use.[9] There are also other methods for such reduction. One of them is a widely used semi-empirical method of the *Buckingham technique in dimensional analysis*[10] (Bridgman P.W., 1922; Qing-Ming Tan, 2011; Gibbings J.C., 2011).

Each system or a process running within it is characterized by a number of technical, economic, aesthetic parameters, etc. Typically, some of them are strongly correlated to each other.[11] This allows defining a surrogate parameter,[12] based on the features of the original. In dealing with many statistical issues, efforts are made to have independent variables for analysis. In *FA* these efforts are manifested by seeking and including correlated variables in one cumulative parameter, thereby obtaining a set of independent features.

CORRELATIONS

The search for correlations between the original features should be a standard procedure for already moderately entangled problematic situations. In many analytical methods,[13] existence of interaction between variables leads to misleading conclusions or complicates the problematic situation to such an extent that proper modelling of studied phenomena is not possible.

A commonly used measure of correlation is the correlation coefficient r, understood here as the *Pearson linear correlation coefficient*.[14] For a two-dimensional random variable (X,Y), the Pearson linear correlation coefficient $r(X,Y)$ can be written as:

$$r(X,Y) = \frac{E\left[(X - EX)\cdot(Y - EY)\right]}{\sqrt{varX \cdot varY}}. \tag{A3.1}$$

where: r – two-dimensional correlation coefficient, X,Y – random variables, EX – expected value of random variable X : $\exists_{P(X \leftarrow a_i)=p_i} EX = \sum_i a_i \cdot p_i$, $varX$ – random variable variance X: $\exists_{P(X \leftarrow a_i)=p_i} varX = \sum_i (a_i - EX)\cdot p_i$

Correlation coefficient r (A3.1) has the following properties:

1) $-1 \leq r \leq 1$
2) $\exists_{X,Y} |r| = 1 \Leftrightarrow Y = a \cdot X + b$
3) $\exists_{X,Y} r = 0 \Rightarrow X \wedge Y$ are uncorrelated

Theorem (3) is not transitive; therefore a lack of correlation between variables X and Y does not lead to the conclusion that they are independent. Indirect proof can be done assuming initially that theorem (3) is true. Demonstration of even only one case of deviation from this assumption will overthrow this thesis. For this purpose we can consider a two-dimensional random variable (X,Y) as having distribution:[15] $P(X=0,Y=0)=P(X=-1,Y=1)=P(X=1,Y=1)=\dfrac{1}{3}\Rightarrow$

$$\forall_{Y=X^2}\ P(X=0)=P(X=-1)=P(X=1)=\frac{1}{3}\wedge P(Y=1)=\frac{2}{3}$$

$$\wedge P(Y=0)=\frac{1}{3}\Rightarrow E(X)=0\cdot\frac{1}{3}-1\cdot\frac{1}{3}+1\cdot\frac{1}{3}=0$$

$$\wedge E(Y)=0\cdot\frac{1}{3}+1\cdot\frac{1}{3}=\frac{2}{3}.$$

From this dependence and equation (A3.1) it follows that

$\forall_{Z=X\cdot\left(Y-\frac{2}{3}\right)}\ r(X,Y)=\dfrac{EZ}{\sigma_X\cdot\sigma_Y}$. Random variable Z has the following distribution:

$P(Z=0)=P\left(Z=-\dfrac{1}{3}\right)=P\left(Z=\dfrac{1}{3}\right)=\dfrac{1}{3}\Rightarrow EZ=0\cdot\dfrac{1}{3}-\dfrac{1}{3}\cdot\dfrac{1}{3}+\dfrac{1}{3}\cdot\dfrac{1}{3}=0$. It follows

that the correlation coefficient $r(X,Y)=0$, although the random variables X and Y are uncorrelated but are dependent q.e.d. (Kłos R., 2012).

THE COEFFICIENT OF DETERMINATION

The correlation coefficient r is associated with the *coefficient of determination* r^2. The smaller the variance is from the sample for residual values $s_r=\dfrac{1}{n}\cdot\sum_{i=1}^{n}(\hat{y}-\bar{y})^2$ around the regression line relative to the total variance[16] from sample

$s=\dfrac{1}{n}\cdot\sum_{i=1}^{n}(y_i-\bar{y})^2$, the better the quality of prediction.[17] In the absence of any dependence between variables X and Y the residual variability s_r to the total variability s ratio would be equal in identity to the unity: $\frac{s_r}{s}\equiv1$. If X and Y were strictly functionally dependent, the residual variability s_r would be equal in identity to zero $s_r=0$ and such a relationship would be $\frac{s_r}{s}\equiv0$. Frequently the ratio is between these values $\frac{s_r}{s}\in[0,1]$. The coefficient of determination value r^2 is defined as:

$$r^2=1-\frac{s_r}{s}=\frac{s-s_r}{s} \tag{A3.2}$$

where: r^2 – determination coefficient, s_r – residual variation from sample, s – total variation from sample.

The coefficient of determination r^2 is regarded as an indicator of how well the model fits the data.[18] It tells what part of the variability associated with random variable Y is explained by the linear regression model.[19]

The total sum of squared deviations $\sum_{i=1}^{n}\left(y_i-\overline{y}\right)^2$ is a sum of the explained with the model, sum of squares $\sum_{i=1}^{n}\left(\hat{y}-\overline{y}\right)^2$, and the variation of the not explained with the model, residual sum of squares $\sum_{i=1}^{n}\varepsilon_i^2$:

$$\sum_{i=1}^{n}\left(y_i-\overline{y}\right)^2 = \sum_{i=1}^{n}\left(\hat{y}_i-\overline{y}\right)^2 + \sum_{i=1}^{n}\varepsilon_i^2 \qquad (A3.3)$$

where: y_i – value from experiment, $\overline{y}$ – global average, $\hat{y}$ – estimator of the value of y_i from the model, ε_i – model error.

Starting from this point of view, the coefficient of determination r^2 can be defined as a ratio of explained variation[20] $\sum_{i=1}^{n}\left(\hat{y}_i-\overline{y}\right)^2$ to total variation[21]

$$\sum_{i=1}^{n}\left(y_i-\overline{y}\right)^2 : \quad r^2 = \frac{\sum_{i=1}^{n}\left(\hat{y}_i-\overline{y}\right)^2}{\sum_{i=1}^{n}\left(y_i-\overline{y}\right)^2} = 1 - \frac{\sum_{i=1}^{n}\varepsilon_i^2}{\sum_{i=1}^{n}\left(y_i-\overline{y}\right)^2}.$$ These relationships are

shown for one point in Figure A3.2.

A higher value of coefficient of determination r^2 means a better correlation. Sometimes, this single point measure can be misleading, also because this coefficient depends on the directional factor of the regression line,[23] regardless of the fact

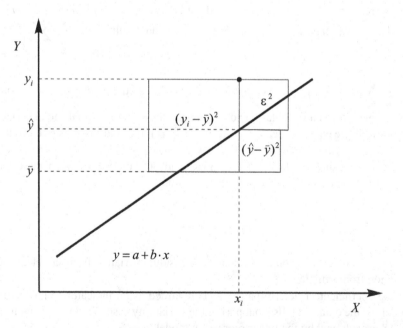

FIGURE A3.2 Breakdown of variation for one point and for linear regression model.[22]

that the variance of the random component does not change. Moreover, adding next point to an existing model is accompanied by the same effect. The improved coefficient of determination $\tilde{r}^2$, where appropriate sums of squares are divided by the number of relevant degrees of freedom,[24] is free of these disadvantages:

$$\tilde{r}^2 = 1 - \frac{S_r}{s} \cdot \frac{V}{V_r} \qquad (A3.4)$$

where: $\tilde{r}^2$ – improved coefficient of determination, V_r – number of degrees of freedom for residual variance $V_r = n - k - 1$, k – number of coefficients of linear combination model $\exists_{x_{1\in\{1,0\}}} \to y = \sum_{i=1}^{k} a_i \cdot x_i$, V – total number of degrees of freedom for variation $V = n - 1$.

This operation takes into account the fact that the r^2 is calculated from a sample and its value is, therefore, overstated when generalizing the results to the overall population. An improved coefficient of determination $\tilde{r}^2$ determines how well the regression equation received would fit another sample having the same number, from the same population.

The improvement $\dfrac{V}{V_r}$ loses its importance for large sample population n because

$$\lim_{n\to\infty} \frac{V_r}{V} = \lim_{n\to\infty} \frac{n-k-1}{n-1} = 1.$$

THE CONVERGENCE COEFFICIENT

The convergence coefficient φ^2 is a point evaluation associated with the coefficient determination r^2 by relation $\varphi^2 = 1 - r^2$. The convergence coefficient φ^2 describes that part of variability of the explained variable, which stems from its dependence on factors other than those included in the model.[25] Like the coefficient of determination r^2 it takes values from the interval $[0,1]$. The closer to zero the value of the convergence coefficient φ^2 is, the better the fit of the model.

STANDARDIZATION

Standardization[26] is most often understood as an activity that involves analyzing systems and processes occurring in them to ensure their functionality, usability, compatibility, safety of operation, limited undue diversity and more by conforming to standards.

The aims of standardization are to ensure functionality of products and services, ensure conformity to standards in production, remove trade barriers, facilitate cooperation in science and technology and so on. Standardization in natural sciences means achieving a direct comparability of quantitative characteristics of objects, phenomena or processes expressed in different physical units. Standardization is achieved by the expression of compared characteristics through a conventionally accepted unit.

In statistics standardization is most often associated with the type of normalization of random variable X, as a result of which the variable achieves the average expected value equal to zero $EX \equiv 0$ and variance equal to one $\sigma^2 = 1$. This operation can also be conveniently performed where arithmetic operations are performed on physical values in order to make variables independent of their units,[27] as sometimes they do not have a physical sense – it is often the case during summation.[28] For standardized variables such a danger does not exist.

Standardization involves subtracting the expected value EX from them and dividing by the standard deviation σ: $\tilde{X} = \dfrac{X - EX}{\sigma}$, where: $\tilde{X}$ – standardized variable.[29] This operation is also required because of the inference method used, since FA is based on rotation of the coordinate system. From a mathematical point of view, this is attained by multiplying values of variables by trigonometric functions of an angle of rotation and adding them to each other.[30] Adding values of different denominations has no physical sense,[31] but for standardized variables such a risk does not exist because they are dimensionless [32] relative values.

COVARIANCE

Generalization of the variance $\sigma_X^2 \equiv var(X)$ for a two-dimensional case is the *covariance cov(X,Y)* describing covariability of two features at the same time: $cov(X,Y) = \sum_i (x_i - EX) \cdot p_i(X) \cdot (y_i - EY) \cdot p_i(Y)$, which is a measure of simultaneous deviations of two values from their average values.[33] It is easy to show now that: $cov(X,X) = var(X)$. For standardized variables $\left(\tilde{X}, \tilde{Y} \right)$, the correlation coefficient $r\left(\tilde{X}, \tilde{Y} \right)$ is equal to covariance $cov(X,Y)$: $\forall_{\tilde{X} \wedge \tilde{Y}} \, r\left(\tilde{X}, \tilde{Y} \right) \equiv cov(X,X)$.

CONCENTRATION ELLIPSE

When there is a sufficiently strong linear correlation between the two-dimensional variables, they will position themselves in a shape similar to the ellipse presented in Figure A3.3.[34]

It follows from this figure that the data presented in it are strongly correlated, and when the coordinate system XY is replaced by the axes of the concentration ellipse $X'Y'$ the measurement points can be characterized using only variable X' without any significant loss of information.

As a first approximation, for the case shown in Figure A3.3, dimensionality of the problem situation can be reduced by rotating the coordinate system by $45°$ and by shifting its center. The covariance matrix defining the interdependence between variables for standardized data transfers into the correlation matrix. The values outside the diagonal of correlation matrix indicate how well the values are correlated with each other.

The covariance/correlation matrix diagonalization $R \rightarrow R'$ illustrates the rotation of the coordinate system by $45°$, which sets the coordinate axes along the principal axes of the scatter plot ellipse (Kłos R., 2007):

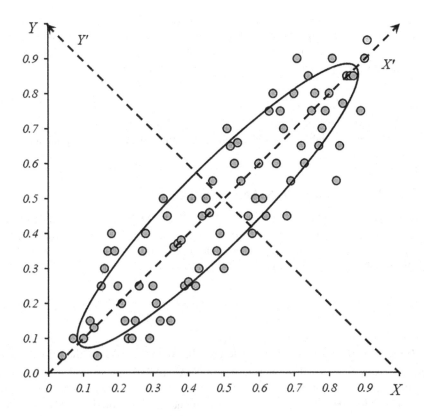

FIGURE A3.3   An example of a scatter plot with a concentration ellipse and its principal axes.

$$\underline{R} = \begin{bmatrix} 1 & r_{xy} \\ r_{xy} & 1 \end{bmatrix} \xrightarrow{obrót\ o\ 45°} \underline{R}' = \begin{bmatrix} a^2 & 0 \\ 0 & b^2 \end{bmatrix} \tag{A3.5}$$

where: $\underline{R}$ – correlation matrix, $\underline{R}'$ – diagonalized correlation matrix, r_{xy} – correlation coefficient, $a^2; b^2$ – constants.

After a more detailed analysis of the problem situation shown in Figure A3.3, it can be concluded that the optimal angle of rotation can vary from 45°. The values of correlation coefficients will form *values of residual correlation*, which will be discussed later.

MAIN COMPONENTS

Once a rotation has been completed, by assumption the total spread contained in the data cannot change,[35] hence from equation (A3.5) $a^2 + b^2 = 1 + 1 = 2$. New variables, resulting from output variables after the rotation of the system, are called *main components*. They have the following characteristics: they are uncorrelated and have average values equal to zero, and they vary only in dispersion.[36] The relationship between the main components of X_1' and the original features X_i is expressed for a particular feature by linear combination:

$$\forall_{k>i>2} \rightarrow X_i = a_{i1} \cdot X_1' + a_{i2} \cdot X_2' + .. + a_{ik} \cdot X_k' \qquad (A3.6)$$

where: X_1' – new variable, X_i – old variable, k – number of main components, a_{ik} – loading factors.

UNIQUE CONTRIBUTION AND UNIQUE VARIATION

The analysis, which takes into account all the main components, does not lead to a reduction of the dimensionality of the problem situation. If, however, some of the main components are omitted, for example, taking into account only two values, equation (A3.6) takes the form:

$$X_i = a_1 \cdot X_1' + a_2 \cdot X_2' + E_i \qquad (A3.7)$$

where: E_i – unique contribution to feature i.

Unique contribution E_i is the difference between the value of the primary feature X_i expressed by all main components[37] X_1' (A3.6) and the value with those omitted, which are considered negligible for the simplified model (Z3.7). The specific unique contribution E_i is a measure of the accuracy of the mapping of the considered system's main features by reducing the number of components. It is easy to submit variables X_1' and X_2' to standardization, after which they are called agents.

Variance of variable X_i from equation (A3.7) can be written down as:

$$V(X_i) = a_1^2 + a_2^2 + V(E_i) \qquad (Z3.8)$$

where: $a_1^2 + a_2^2$ – resources of common variability, $V(E_i)$ – unique variance.

SIMPLIFICATION OF FACTOR STRUCTURE

After performing the prior operations, the relationship between the measured characteristics and factors may still be difficult to interpret. To simplify their analysis a simplification strategy for factor structure was developed. The structure obtained can be subjected to a rotation again so that some factor loadings a_{ik} should reach high values and others should have values close to zero – creation of the so-called *simple structure*. It is convenient to implement these strategies using appropriate computer programs.[38] The most common strategy is the *varimax rotation*, which depends on further maximizing the variance for the characteristics which in the factor structure have the greatest value. After performing such a rotation, the primary variables arrange themselves in the proximity of the selected factors axes in the space.

ROTATION

Figure A3.4 shows examples of the results of N observations for two standard variables.

Because of the applied standardization, the center of the coordinate system was selected to be at the point corresponding to the average samples. Each point in the graph represents a feature vector measured in the experiment.[39] In this case, the line U_A cannot be adjusted to the cloud of points,[40] which, together with the perpendicular axis U_B, create a new system of coordinates, rotated in relation to the primary system by an angle α. It should be noted that the rotation of the system of coordinates was made in such a way that the greatest variability is achieved along the axis U_A. This type of rotation is called variance maximizing[41] because it maximizes the variance along the axis U_A and minimizes it along the axis U_B. Because the system is orthogonal $U_A \perp U_B$, the new variables become independent. The relationship between the new and old variables can be found from the projection of vectors $\hat{X}_{A,i}$ and $\hat{X}_{B,i}$[42] 436 onto the UA and UB axes. They will create the following linear combinations:

$$\begin{cases} U_A = \hat{X}_A \cdot \cos\alpha + \hat{X}_B \cdot \cos\left(\dfrac{\pi}{2} - \alpha\right) \\ U_B = \hat{X}_A \cdot \cos\left(\dfrac{\pi}{2} + \alpha\right) + \hat{X}_B \cdot \cos\alpha \end{cases} \Rightarrow \begin{cases} U_A = \hat{X}_A \cdot \cos\alpha + \hat{X}_B \cdot \sin\alpha \\ U_B = \hat{X}_A \cdot \sin\alpha + \hat{X}_B \cdot \cos\alpha \end{cases} \quad \text{(A3.9)}$$

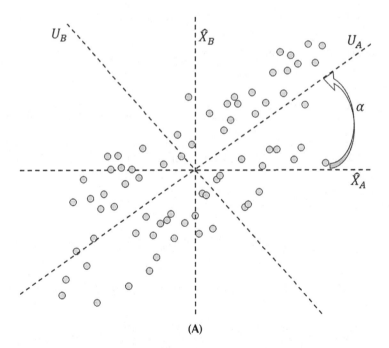

(A)

FIGURE A3.4 Example of diagram of N observations for two standard variables.

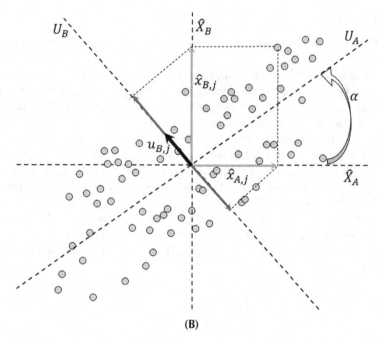

(B)

FIGURE A3.4 (Continued)

By defining the rotation matrix as: $\underline{Q} = \begin{vmatrix} \cos\alpha & \sin\alpha \\ -\sin\alpha & \cos\alpha \end{vmatrix}$ the rotation operation

can be written down in the matrix form: $\underline{U}^T = \underline{Q} \cdot \underline{\hat{X}}^T$ where $\underline{\hat{X}} = \begin{vmatrix} \hat{X}_A & \hat{X}_B \end{vmatrix}$ and

$\underline{U} = \begin{vmatrix} U_A & U_B \end{vmatrix}$. Vector $\underline{U}$ is called the *principal component* or the *principal factor*.

Table A3.1 presents the data generated for a linear relationship $y = 2x + 3$, dispersed by random numbers generator (Figure A3.5a). The factor loadings for standardized input data and the relationship between the new and old variables are presented in the form of matrix:

$$\begin{bmatrix} \hat{x} \\ \hat{y} \end{bmatrix} \cong \begin{bmatrix} 0.9981 & 0.0611 \\ 0.9981 & -0.0611 \end{bmatrix} \cdot \begin{bmatrix} factor1 \\ factor2 \end{bmatrix} \tag{A3.10}$$

*Factor*1 of the equations (A3.10) is associated with the variability explained, while *Factor*2 with noise. The inverse transformation can be written using the following formula:

$$\begin{bmatrix} factor1 \\ factor2 \end{bmatrix} \cong \begin{bmatrix} 0.5009 & 0.0611 \\ 8.762 & -8.1762 \end{bmatrix} \cdot \begin{bmatrix} \hat{x} \\ \hat{y} \end{bmatrix} \tag{A3.11}$$

TABLE A3.1

Output for dimensional analysis for evenly disturbed linear function: $y = 2x + 3$.

x	y	x	y	x	y	x	y
−3.0000	−3.2143	−1.4000	−0.0392	0.2000	4.0717	1.8000	7.1723
−2.9000	−2.1459	−1.3000	0.4830	0.3000	3.9542	1.9000	7.2405
−2.8000	−2.3810	−1.2000	1.0558	0.4000	3.1750	2.0000	7.0905
−2.7000	−1.9818	−1.1000	0.4438	0.5000	4.2329	2.1000	6.9062
−2.6000	−2.3258	−1.0000	1.4652	0.6000	4.3840	2.2000	7.5364
−2.5000	−2.8843	−0.9000	0.7683	0.7000	3.8488	2.3000	7.3930
−24000	−1.3225	−0.8000	1.5112	0.8000	5.2760	2.4000	8.3008
−2.3000	−1.8526	−0.7000	2.0381	0.9000	5.1808	2.5000	7.5313
−2.2000	−1.0711	−0.6000	1.5111	1.0000	5.0900	2.6000	7.7800
−2.1000	−0.9711	−0.5000	1.9643	1.1000	4.7767	2.7000	8.7451
−2.0000	−1.1825	−0.4000	3.0849	1.2000	6.0682	2.8000	8.3808
−1.9000	−1.4046	−0.3000	2.0150	1.3000	5.8643	2.9000	8.4622
−1.8000	−0.4872	−0.2000	2.1014	1.4000	6.2426	3.0000	9.7035
−1.7000	−0.2974	−0.1000	2.6650	1.5000	5.9581		
−1.6000	−0.1716	0.0000	3.1370	1.6000	5.7558		
−1.5000	0.3845	0.1000	4.0051	1.7000	6.2429		

Taking into account only *factor1*, the results of the transformation can be written as:

$$\begin{cases} factor1 = 0.5009 \cdot \hat{x} + 0.5009 \cdot \hat{y} \\ \hat{x} = 0.9981 \cdot factor1 \\ \hat{y} = 0.9981 \cdot factor1 \end{cases} \qquad (A3.12)$$

The results of these transformations are in Figure A3.5b, showing that the use of *FA* gives satisfactory results.[43]

RESIDUAL CORRELATION

In the spatial analysis of principal components, once a selected number of factors have been isolated, there still remains a variation around the axis along which the variance has the maximum magnitude. It can be neglected or an attempt can be made to isolate another orthogonal principal component. The basic criterion here will be the percentage of phenomenon explanation. As mentioned earlier, an analysis in which all possible principal components are taken into account does not lead to a reduction in dimensionality of the problem. It must be decided how much of the information[44] may be lost without harm to the problem investigated. The new structure of independent variables can be further simplified and optimized. Such operations should be computer-aided as theoretically simple linear algebra calculations are burdensome to perform.

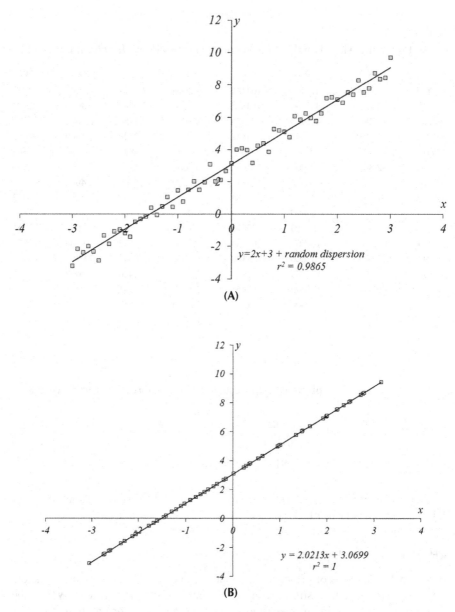

FIGURE A3.5 Example of *FA* for slightly evenly impaired linear function $y = 2x + 3$: (a) standardized output data, (b) standardized data after removal of the dispersing agent.

PRINCIPAL COMPONENTS

An example of variables strongly correlated into one factor shows the idea of minimizing the number of variables[45] and replacing them with the principal component analysis. In addition, the principal components are orthogonal and thus they are uncorrelated and independent, their average value is zero, and they differ only in the dispersion, which is the basic condition to be met when certain methods of analysis of measurements results are applied.

SPIROMETRY

Table 7.1 in Chapter 7 shows the results of spirometry tests performed after all the 17 experimental oxygen dives: VC – vital lung capacity [dm^3], FEV_1 – first second forced expiratory volume [dm^3], FVC – forced vital lung capacity [dm^3], PEF – peak expiratory flow [$dm^3 \cdot s^{-1}$], $FEF25–75$ – forced exhaust flow counted between 25% and 75% of vital lung capacity [$dm^3 \cdot s^{-1}$], PIF – peak inspiratory flow [$dm^3 \cdot s^{-1}$], before, immediately after and after 1 h of oxygen exposure. These results were subjected to factor analysis.

The analysis of the correlation coefficients between the respiratory parameters shows a strong correlation between the values of forced parameters (Table A3.2 and Figure A3.6).

Leaving two factors from the researched case for further analysis, it is necessary to check how they reproduce the correlation matrix by calculating residual correlations. They are the difference between the correlation coefficients in output matrix and the correlations calculated on the basis of the factor values (Table A3.3).

Table A3.3 contains two factors from the researched case left for further analysis that should be investigated to find out how they reproduce the correlation matrix, by calculating residual correlations. They are a difference between the correlation coefficients in output matrix and the correlations calculated on the basis of the factor values (Table A3.4).

It can be seen that the greatest difference is 0.20, and it differs from the other differences in that they are small – the description was considered satisfactory.

The factorial structure in Figure A3.7 is sufficiently clear and the use of an additional rotation does not make any improvement.

TABLE A3.2
Correlation coefficients for spirometry measurements.

Parameter	VC	FEV₁	FVC	PEF	FEF 25–75	PIF
VC	1.00	0.46	0.23	−0.21	0.59	−0.43
FEV₁		1.00	0.88	0.05	0.93	−0.29
FVC			1.00	0.07	0.65	−0.25
PEF				1.00	-0.02	0.18
FEF25–75					1.00	−0.33

VC – vital lung capacity [dm^3], FEV_1 – first second forced expiratory volume [dm^3], FVC – forced vital lung capacity [dm^3], PEF – peak expiratory flow $\left[dm^3 \cdot s^{-1}\right]$, FEF25–75 – forced exhaust flow counted between 25% and 75% of vital lung capacity $\left[dm^3 \cdot s^{-1}\right]$, PIF – peak inspiratory flow $\left[dm^3 \cdot s^{-1}\right]$.

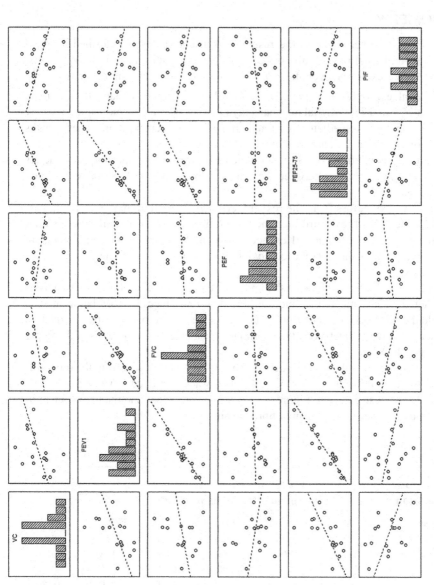

FIGURE A3.6 Qualitative analysis of linear relationships between the measured spirometry parameters: VC – vital lung capacity [dm^3], FEV_1 – first second forced expiratory volume [dm^3], FVC – forced vital lung capacity [dm^3], PEF – peak expiratory flow [$dm^3 \cdot s^{-1}$], $FEF25-75$ – forced exhaust flow counted between 25% and 75% of vital lung capacity [$dm^3 \cdot s^{-1}$], PIF – peak inspiratory flow [$dm^3 \cdot s^{-1}$].

TABLE A3.3
Factor loadings before rotation.

Variable	Factor1 : inspiration parameters	Factor2 : expiration parameters
VC	-0.66	0.46
FEV_1	-0.95	-0.26
FVC	-0.81	-0.36
PEF	0.06	-0.78
FEF25–75	-0.93	-0.08
PIF	0.51	-0.50
Explained variation	3.12	1.28
Contribution to variance	52%	21%

Factor No	Eigenvalue	Contribution to variance	Cumulated eigenvalue	Cumulated Contribution to variance
1	3.12	52.1%	3.12	52.1%
2	1.28	21.4%	4.40	73.5%

TABLE A3.4
The residual correlation coefficients for the spirometry measurements.

Parameter	VC	FEV_1	FVC	PEF	FEF 25–75	PIF
VC	0.35	-0.05	-0.14	0.20	0.02	0.14
FEV_1		0.03	0.01	-0.09	0.03	0.06
FVC			0.22	-0.16	-0.13	-0.02
PEF				0.38	-0.02	-0.25
FEF25–75					0.13	0.11

As can be seen, one of the factors is strongly associated with the forced inspiratory parameters: FEV_1, FVC, FEF25–75 and the other with the peak expiratory flow PEF. The peak inspiratory flow PIF and vital capacity of lungs VC make up to half of each of these two factors (Table A3.3).

SUMMARY

The obtained results of spirometry parameters create a relatively small database, which is characterized by a large scatter.[46] In such a case the confidence interval is too wide, which makes it difficult to draw effective statistical conclusions (Figure A3.8 and Table 7.2).

However, the results of the factor analysis run here show a strong correlation for the spirometry parameters for stress exercises. It can be noticed that the parameters listed in the set of selected parameters cannot be added to any group of vital capacity VC and peak inspiratory flow PIF. In addition, their shares are distributed in half to two factors possible to isolate. Therefore, in the set of standard spirometry parameters it is better to use relative values – criterial numbers, built from various

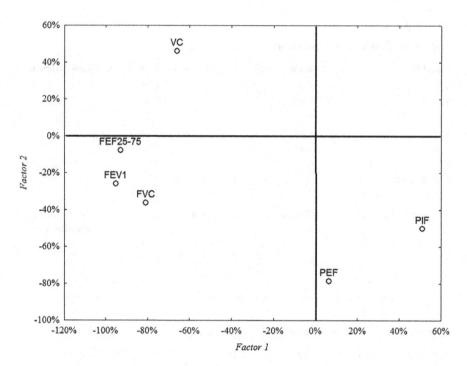

FIGURE A3.7   Factor structure for the spirometry measurements with no rotation.

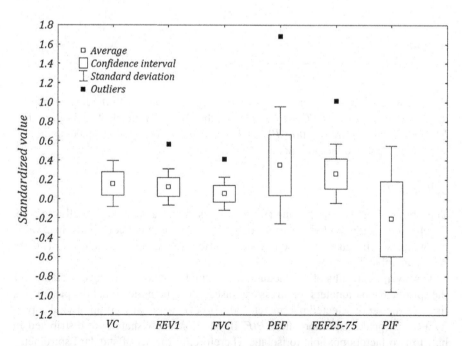

FIGURE A3.8   Standardized measurement of spirometry parameters: squares represent average values without outliers, frames represent standard deviation, runners represent expanded uncertainties for the probability $P = 0.95$, circles represent protruding points.

respiratory parameters. Indirect proof is the *Tiffeneau indicator* frequently used in medical diagnostics, which determines the ratio of first second in the forced expiratory volume FEV_1 to the vital capacity VC.

NOTES

1 Without modelling, we are unable to analyze the reality around us as it is too complicated. In many cases there is no need to use complex models, since it is possible to analyze a system or process that runs in it, based on the simpler versions. Degree of simplification and the choice of the parameters of knowledge that will be reproduced in the model depending on the purpose for which it is created.

2 System of assumptions, concepts, relationships, etc.

3 Physical, biological, economic, etc.

4 The notion of *mathematical model* also includes instruments used to solve or interpret interrelationships in a tested segment of reality, different concepts and mathematical theories, etc.

5 *Paradigm* (Latin *paradigm* – example, pattern) is a basic axiom in a given field of knowledge.

6 Formerly called *data digging*.

7 Reduction in problem situation dimensionality.

8 For example, a number of technical parameters of a system or process.

9 For example, to allow for an unambiguous classification of the device as a high or low class – like in dimensional analysis, using the *Reynolds number*, flow types can be divided into three groups: laminar, transitional and turbulent.

10 Chemical, process and sanitary engineering, etc.

11 For example, fuel consumption with vehicle weight and with achieved acceleration.

12 Which would allow, for example, the classification of system, understood as an issue to be decided in the problem case.

13 Not only statistical.

14 Another popular measure of correlations is *Spearman's rank correlation coefficient* using the same relationship as the Pearson coefficient with the difference that the data consist of ordinal ranks representing ordinal values for data, rather than of the results obtained. Interpretation of Spearman's rank correlation coefficient is the same, but is not narrowed to linear regression. This factor is used for testing nonparametric hypotheses, determining the correlation regardless of the statistical distribution, determining the regression, of any type, etc.

15 Random variables X and Y are probabilistically dependent, when: and are functionally dependent, because: $P(X = -1, Y = 1) = \frac{1}{3} \neq P(X = -1) \cdot P(Y = 1) = \frac{1}{3} \cdot \frac{2}{3} = \frac{2}{9}$.

16 Total variability.

17 Formulating a judgment.

18 r^2 close to 1 indicates that almost all of the variation of the dependent variable can be explained by the independent variables included in the model.

19 For example, the value of the coefficient of determination $r^2 = 0.4$ says that the variance of Y values around the regression line is 10.4 of original variance, because on the basis of the definition (A3.2) s_r value $s_r = (1 - r^2) \cdot \sigma$; in other words, 40% of the original variation Y is explained by the regression, and 60% remains in the residual variability.

20 Sum of model deviation squares from the average value.

21 The total sum of the squared experimental deviations value from the average value.

22 As one can see from the figure, one-point fields do not add up: $(y_i - \bar{y})^2 \neq (\tilde{y}_i - \bar{y})^2 + \varepsilon_i^2$.

23 The greater the angle of inclination of the regression line, the greater is determination coefficient r^2.

24 Improved coefficient of determination $\hat{r}^2$ is always smaller from the determination coefficient r^2: $\hat{r}^2 < r^2$.

25 Defines which part of changeability of variable has not been explained through a model.

26 *Standardization* is the development and implementation of standards that are an average of norms, the average type of model, a product that meets the defined requirements, pattern, etc.

27 Physical dimension.

28 Sometimes physical quantities in the same units may not be additive, for example, mixing one liter of water with one liter of ethyl alcohol does not lead to formation of two liters of a solutions of 50% ; there will be less of it, and it will be more concentrated.

29 It is noticeable that:

$$E\tilde{X} = E\left(\frac{X - EX}{\sigma}\right) = \frac{1}{\sigma} \cdot \left[E(X - EX)\right] = \frac{1}{\sigma} \cdot \left[EX - E(EX)\right] = \frac{1}{\sigma} \cdot \left[EX - EX\right] = 0.$$ The

standard deviation for the case under consideration here is a fixed value, hence the variance

$D^2\tilde{X} = D^2\left(\frac{X-EX}{\sigma}\right)$ can be written further as:

$D^2\tilde{X} = \frac{1}{\sigma^2} \cdot D^2(X - EX) = \frac{1}{\sigma^2}[D^2X - D^2(EX)]$. Since $D^2(EX)$ is the variance of the

expected value EX, which is a fixed value, it is equivalent to zero by definition $D^2(EX) = 0$.

The value of the variance D^2X by definition is equal to the standard deviation of $D^2X = \sigma^2$. Thus, ultimately, you can write: $D^2\tilde{X} = \frac{\sigma^2}{\sigma^2} = 1$. Summarizing, the standardized variable is the unitless value with zero expected value and variance equal to

$$\forall_{\tilde{X} = \frac{X - EX}{\sigma}} E\tilde{X} = 0 \wedge D^2\tilde{X} = 1.$$

30 That is, forming their linear combinations.

31 They are often not additive, as their sum does not have a physical interpretation – for example, it makes no sense to add mass and volume.

32 Their dimension is identically equal to unity $\left[\tilde{X}\right] \equiv 1$.

33 For linear regression, covariance is correlated with the slope of a straight line a:

$$\forall_{Y = a \cdot X + b} \, a = \frac{cov(X,Y)}{var(X)} \quad .$$

34 Generalizing on a larger number of dimensions we talk of hyper-ellipse concentration.

35 Total data scatter equals the sum of all the variances.

36 Variance.

37 New variables.

38 For example, STATISTICA.

39 That is, the measurement result – for example Figure A3.3 sample point with coordinates $\left(\hat{X}_{A,i}; \hat{X}_{B,i}\right)$.

40 For example, linear regression.

41 *Varimax.*

42 For ith measurement.

43 Showing resistance of the method to the applied distortion, although obtained parameters of the straight line differ from the original by 75%.

44 Variability.

45 Therefore the reduction in dimensionality of the problem situation.

46 As one could have expected for biological measurements.

REFERENCES

Bastlevsky AT. 2010. *Statistical Factor Analysis and Related Methods Theory and Applications.* New York: Wiley-Interscience. ISBN 978-04-715708-20.

Bridgman PW. 1922. *Dimensional Analysis.* New Haven: Yale University Press.

Gibbings JC. 2011. *Dimensional Analysis.* London: Springer-Verlag Ltd. ISBN 978-1-84996-316-9; e-ISBN 978-1-84996-317-6. http://doi.org/10.1007/978-1-84996-317-6.

Hinkelmann K & Kempthorne O. 2005. *Design and Analysis of Experiments. Volume 2: Advanced Experimental Design.* Hoboken: John Wiley & Sons, Inc. ISBN 0-471-55177-5.

Kłos R. 2007. *Zastosowanie metod statystycznych w technice nurkowej – Skrypt.* Gdynia: Polskie Towarzystwo Medycyny i Techniki Hiperbarycznej. ISBN 978-83-924989-26.

Tan Q-M. 2011. *Dimensional Analysis with Case Studies in Mechanics.* Berlin: Springer-Verlag. ISBN 978-3-642-19233-3; e-ISBN 978-3-642-19234-0. http://doi.org/10.1007/978-3-642-19234-0.

Index

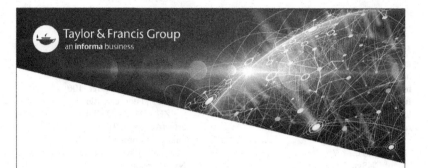

Printed in the United States
by Baker & Taylor Publisher Services